Modern Biotechnology in PLANT BREEDING

Modern Biotechnology in PLANT BREEDING

Editor
Dharamvir Hota

2007
Gene-Tech Books
New Delhi - 110 002

ISBN 81-89729-18-7

Published by : GENE-TECH BOOKS
4762-63/23, Ansari Road, Darya Ganj,
NEW DELHI - 110 002
Phone: 41562849
e-mail: genetechbooks@yahoo.co.in

Printed at : Tarun Offset Printers
Delhi

PRINTED IN INDIA

PREFACE

Recent progress in explaining the science behind biological processes has significantly enhanced our ability to exploit them for the benefit of mankind and the environment. The evolution of plant breeding is a classic example of how improved biological understanding has been adapted to provide more effective methods of meeting the demands of a changing world. All forms of plant breeding involve the improvement of crops by selection. Since man began farming he has selected seeds from the best plants for the next generation. Over the centuries, this selection process has become more scientific, bringing major improvements in the yield, quality and diversity of crops.

Modern biotechnology is the latest stage in the development of plant breeding technology. The tools of the biotechnologist increased the speed and precision of plant breeding techniques and widened the choice of characters for selection. Modern biotechnology allows crop improvement to take place at the level of individual genes. Techniques such as genetic modification enable plant breeders to identify the single gene responsible for a particular trait, and insert, delete or modify it in a plant variety. This extends the range of characters at the breeder's disposal, and enables specific genes to be expressed in a crop plant without the uptake of unwanted characteristics which must then be selected out.

This book provides an overview of the applications of modern biotechnology in plant breeding. It will certainly meets with interests and needs of up to date information of plant breeders, may they be advanced students, practitioners, teachers or researchers.

Editor

Contents

1

INTRODUCTION

The cultivation of plants is an important aspect of today's society in many ways. Not only do plants supply with a major food resource and flow of nutrition but they are also an important source of chemicals and other non-food products such as drugs, oils, latex, pigments, and resins. Because of the high value of plants, farmers are constantly trying to increase the yield and quality of their products by using more effective production techniques. The physical appearance of plants is determined by genetic factors that are inherited from parental crops. The natural transfer of genes can be altered using many genetic techniques such as selection or genetic engineering.

One of the important goals of plant breeding is maintaining genetic diversity. This can be achieved through selection and hybridization. More recently, new techniques have evolved to allow to grow crops artificially in culture from plant parts to produce whole plants. Modern science has allowed to get away from the traditional method of plant breeding to more artificial asexual methods. This allows plants to be produced using individual, or groups of cells, which grow into full plants. Genetics has also allowed to modify crops in a more scientific way to select for traits that are more desired by society.

Plant improvement by man began many thousands of years ago. Pre-agricultural man learned that seeds put into the ground at a certain time of the year produced similar seed-producing plants. This was the beginning of domestication of plants and led to the production of the first crops. Soon after, it was discovered that some crops of the same species grew better or tasted better than others, which became the starting point of selection breeding.

Later in the nineteenth century, Charles Darwin brought to light his theory of Natural Selection, which he described as "survival of the fittest". It was this statement that showed how selection occurred naturally with the environment selecting which plants would grow in specific parts of the world. Selection by farmers was an extension of this process. Farmers were able to further select for plants that not only grew well, but had better taste and looking more pleasing to the eye.

Later, selection was further applied to find crops that produced the highest yield with the best possible quality. In order for one to undergo selection, the breeder must have a large population of seeds to select from. This population should contain several different genetic types that are well adapted to the seasonal changes in the environmental conditions they are grown. These populations are called landraces and are generally found in regions where older plants have been growing for a long time.

Natural selection has already occurred leaving only those crops that have been able to adapt to their environment. Landraces are generally found in areas where cross-breeding has been able to occur giving rise to many new genetic varieties with new genotypes constantly being introduced into the environment. Once an appropriate population is found, one can begin the process of selecting plants by removing those plants with undesired traits. This allows plants with the most desired traits to survive and give rise to a new line of crop that has been selectively chosen for its specific traits.

Hybrid Breeding

Another method of selection in plants occurs when two different pure lines are crossed giving rise to what is called a hybrid. Hybrids have the genetic basis of each of the traits of the parents. In this way, one can obtain plants with the genetic basis for many desirable traits exhibited by each of the parents. By further selection of the offspring, one can select for desired traits

Although selection has been fairly effective over the years and is still used in some places today, it has not been able to keep up to the growing need for plants and plant products. Selection takes time and many things have to be considered, especially environmental factors that cannot always be controlled. It is for this reason that new and more effective methods of plant breeding have been discovered to give rise to an increased number of plants over a shorter period of time.

With a basic understanding of how genetics plays a role in determining the appearance and shape, as well as growth responses of a plant, geneticists have been able to further their knowledge and apply new techniques for growing plants. With a foot in the door as to what gives rise to different traits, geneticists were able to take this one step further by modifying the genetics of a certain plant artificially using an asexual technique to grow these plants. This enables plant scientists to select for the traits that were most suitable for the crops they are growing without having to go through the long and tedious procedures of growing the crops and selecting for the traits that are within that population.

New techniques involve growing whole plants from single cells artificially in cultures that contain all the required nutrients and factors involved in cell growth. This technique is referred to as plant cell culture. Furthermore, scientists have been able to clone specific genes from one species and insert them into plants to introduce new traits that may benefit the plant species. This procedure is known as transgenics and gives rise to genetically modified crops. Furthermore, selection was put into operation to find crops that could withstand certain pests and environmental conditions, such as drought or freezing temperatures.

Although plant science is a relatively new field of interest, there has been a great deal of new technology in this field. Plant scientists have been able to overcome the slow reproductive phase involved in growing plants and have been able to produce plants that can grow in half the time with much better results. In order to make up for the growing demand of crops in today's society, working with geneticists has enabled this to occur. Although old methods of plant breeding are still practiced today in many parts of the world, especially in developing countries, many developed countries have switched to using these new technologies. However, as for anything there are limitations and concerns for both methods.

The more primitive methods of selection take a long time and do not always provide the desired crops. In addition, when a farmer does finally achieve his desired plant product, industry has changed and a new trait is preferred over the previous one. In a constantly growing and changing world, it seems that the farmer can never get ahead. Also, changing environmental conditions can lead to decreased crop yield from year to year and possibly even cause the farmer to lose entire crops. Other disasters, such as fire and disease can lend its destruction to crops causing much devastation.

In the more modern methods of plant growth, crops can be grown in artificial medium requiring less landmass to produce large amounts of crops in less time. Although this seems like a great alternative to the earlier methods, it can also be devastating. By growing plants at a faster rate, may be losing the essential vitamins and nutrients that are important for us. Incorporating foreign genes has been suggested to lead to allergies in many people as well as affect the food chain of other organisms.

Transgenics is still a relatively new field and no concrete evidence for any of this exists but it is growing concern. Plant science is a growing field that is constantly evolving to promote changes that benefit both the industry and accommodate the growing populations that are facing. With the many innovative ideas and constant findings, plant science has provided many great additions to society.

Genetic Basis of Crop Improvement

Agricultural technology is based on the domestication of wild plants to create the crops that we have come to depend upon. Humans invented agriculture approximately 10,000 years ago when they began to harvest and cultivate specific plants to produce food. The improved plant traits selected by early agriculturalists were transmitted genetically to succeeding generations of plants. Domestication of corn by prehistoric agriculturalists has modified the plant to such an extent that it hardly resembles the wild teosinte plants from which it was originally selected.

The majority of the altered traits associated with domestication of crop plants were already accomplished by the time of historic agricultural civilisations such as the Egyptian, Chinese, or Mayan. In addition, these primitive crop cultivars, also known as land races, were adapted to local growing conditions and preferences, and therefore remained genetically diverse for traits such as product qualities, stress tolerance, disease resistance, and yield stability. The domesticated plants on which agriculture is based have resulted from the genetic modification of wild plants through thousands of years of gradual selection.

Scientific studies of genetics began around 1900 based on the work of Gregor Mendel. We now know that genes, the hereditary units, are composed of DNA (deoxyribonucleic acid) bundled together in the form of chromosomes within the nucleus of every living cell of an organism. All of the genetic instructions that determine an organism's characteristics are "packaged" in the genes contained in the chromosomes. These instructions are coded in the specific sequences of the four components of the DNA molecule: adenine (A), thymine (T), cytosine (c), and guanine (G).

The biological systems within the cell convert these instructions into proteins, with the DNA sequences coding for the corresponding protein. The activities of these proteins execute all the functions that determine the characteristics of the organism.

Organisms that reproduce sexually contain two copies of every chromosome. During reproduction, each parent randomly transmits only one chromosome from each of its chromosome pairs to its offspring; therefore, different progeny of the same parents can vary in their genetic composition. By selecting superior individuals from each generation and using them as parents for the next generation, agriculturalists gradually developed crop varieties having more desirable traits and fewer undesirable traits.

An understanding of genetic principles and their application to plant breeding technology has greatly accelerated the rate of improvement of crop plants. Breeders maintain a continual search for novel genetic combinations from which to select plants with superior traits, such as crop quality, yield, regional performance, and tolerance to pests and diseases.

Breeders often make sexual crosses between diverse genotypes to produce new combinations of genetic traits, which then result in diverse phenotypes, or observable morphological or quality traits in the progeny plants. A primary source of genetic variation is the wide array of germplasm within each crop species and closely related wild species that are capable of interbreeding. For many crops, breeders have relied heavily on the introduction of genes from closely related wild plants to increase genetic variation in the crops.

Hybridization between a crop plant and a related wild species enables valuable genes from the wild species to be used for genetic improvement of the crop plant. Virtually all modern crop varieties incorporate resistances to fungal, bacterial, and viral diseases that have been introduced through wide crosses between domesticated varieties and related wild species. Genetic variation can also be increased by inducing mutations, or changes in the DNA sequences of the plants. Since the 1950s, over 2,000 crop varieties have been developed by inducing mutations to randomly alter genetic traits and then selecting among the progeny for improved types.

Specific hybridization methods are also used to increase crop uniformity and yield. Scientists discovered in the 1920s that if some plants, such as maize, were continually crossed with themselves, their vigor would quickly decline over several generations. However, if two distinct inbred plants were crossed together, the resulting progeny often exceeded both parents in vigor and yield. (called *F1 hybrids*) This phenomenon, known as hybrid vigor or heterosis, is the basis of a large hybrid seed industry for maize, sorghum, sunflower, and many vegetables.

Producing hybrid seeds requires genetic manipulations to develop the inbred parents. A specific cross is then made between the two parents to produce the F1 seed that is sold to the farmer. Plants growing from those seeds exhibit the desirable hybrid vigor to produce the commercial crop, but the subsequent generation (F2) loses this advantage due to the random assortment of the traits from the two parents among the individual progeny, in accordance with the principles of genetics. This means that farmers must purchase new seed each year in order to receive the higher yield and other benefits that the hybrid crop provides.

Regardless of the method by which genetic variation is introduced or managed, diverse progeny must be tested extensively to identify the superior plants that will be propagated and used for commercial production of food or fiber. Every year breeders test large populations of plants using various techniques depending on the crop species and the mode of inheritance for the desired traits. Generally, only a very small percentage of the plants tested exhibit improved traits and are therefore retained to be parents for the next generation. In combination with improved agronomic practices, genetic improvement of crop varieties has led to dramatic increases in crop yield and quality.

It is generally estimated that about half of the total increases in crop yield in the past 50 years are the result of genetic improvements, the other half being due to improved management techniques.

Both public institutions and private companies are involved in crop genetic improvement. Private investment in plant breeding has been stimulated by legal mechanisms to allow the developer to retain commercial rights to the sale of the resulting variety. Since 1930, the U.S. Plant Patent Act has allowed breeders to obtain patent protection for plant varieties developed in many ornamental, tree, and vine crops that are propagated vegetatively. The Plant Variety Protection Act of 1970 established breeders' rights for development of plant varieties propagated by seed, and general utility patents can also be awarded for improved plants. These laws are intended to prevent unauthorised propagation and sale of protected varieties, much as copyright laws prevent copying of books or music without a license from the owner. In addition, F1 hybrid varieties offer a biological protection method because they cannot be propagated by simply saving seed from the commercial crop, as is possible with many self-pollinated varieties such as soybean, wheat, or cotton.

Tool for Plant Improvement

The plant breeding methods described above have enabled remarkable improvements to be made in the yield and quality of modern crops. Additional tools are also used for genetic enhancement and propagation of crop plants. As with plant breeding, these methods have been used for many years to develop crop plants that better suit needs.

Grafting and Tissue Culture

Grafting of tissues from two different varieties of a plant species has been used since ancient times in woody tree and vine crops such as citrus, peaches, walnuts, grapes, and ornamental trees.

Surgically cutting a scion or bud from one variety and grafting it onto a rootstock from a different genetic variety is commonly used to enhance the disease resistance, productivity, and growth habit of these perennial crops. Superior varieties, often developed through chance or induced mutations as well as through sexual crosses, can be rapidly and uniformly propagated by grafting buds onto the rootstocks of other varieties.

Tissue culture has been used in crop improvement since the 1940s. In the simplest cases, this refers to culturing embryos or small plants in the laboratory on specific nutrient media until they can be moved into soil. The tiny growing tips of plants *(meristems)* can also be grown in culture to produce entire plants. In addition, plants have the unique property of being able to regenerate an entire plant from a single cell *(totipotency)*. Under carefully controlled conditions, tissues can be taken from a plant, separated into individual cells that are grown in the laboratory as callus, then induced to develop back into whole plants. These techniques have been used in a number of ways in crop propagation and improvement.

Micropropagation is the production of multiple copies of a single plant using tissue culture techniques. Often, the tissue used is the meristem, where new leaves and stems are produced. Each growing tip on a plant can be excised and grown into a complete new plant. In many cases, viruses that infect a plant are not present in the growing tip, so viruses that have infected the parent plant are eliminated from plants produced by micropropagation. Potatoes are normally propagated by planting the buds, or "eyes," present on the tubers, and garlic is propagated by planting cloves from last year's crop. This method of propagation allows viruses to be transmitted to the new crop each year, resulting in diseased plants and reduced yields.

Micropropagation of potato and garlic plants in the laboratory can eliminate virus diseases and ensure that each new crop is planted with virus-free materials, greatly increasing yields. Micropropagation has been the starting point for each new cycle of strawberry plants grown in California for over 20 years both in commercial fields and home gardens.

Micropropagation is also used in asparagus and other crops to multiply individual superior plants to create parent plants used in production of hybrid seed. Many ornamental plants (e.g. orchids and gerbera) are now almost exclusively propagated in this way. Embryo culture has been used to rescue hybrid plants from wide crosses, which often fail to produce mature viable seeds. In these cases the immature embryo tissue can be removed from the developing seeds and cultured in the laboratory to produce the hybrid plants.

Embryo culture enables the breeder to successfully make wide crosses with a greater number of related species of wild plants and have access to a much wider range of genes that can be used for genetic improvement of crop plants. Wide crosses and embryo culture have been valuable tools, especially for the transfer of disease resistance genes from wild relatives into crop plants.

Anther culture is a specific application of tissue culture used in the production of F1 hybrid varieties. The first step is to develop inbred parent lines by repeated self-pollination. This can be a very slow process in some crops, such as onions or carrots, which normally require 2 years to flower. However, by culturing pollen grains in the laboratory, *haploid* plants that contain only one copy of each chromosome can be produced.

These plants can then be induced to double their chromosome number by a chemical treatment, quickly resulting in plants that have two identical sets of chromosomes, or are completely inbred. This procedure can dramatically reduce the time required to develop inbred parents for breeding of F1 hybrid varieties and facilitates the selection of recessive traits. (*homozygous*).

Cell culture (*callus*) can also be used to produce multiple embryos that can be grown into plants. In this way, a virtually unlimited number of plants can be propagated from a single superior plant. However, *somaclonal variation* somewhat limits the application of cell culture for plant propagation. Somaclonal variation is genetic variation that occurs in a small fraction of cells in a callus culture.

Most plants regenerated from callus are genetically identical to the original plant from which the callus was derived, but a small fraction of the resulting plants may exhibit genetic changes. While this is a problem for the clonal *propagation* of identical plants, it has been exploited as a new method to increase genetic variation. Carrot and pepper varieties with better eating quality have been selected among plants derived from callus culture. Protoplast fusion is another method for producing new genetic variation in the laboratory.

A protoplast is a plant cell from which the rigid outer cell wall has been removed using enzymes. Under certain conditions, protoplasts derived from genetically different plants can be fused much like two soap bubbles, allowing their DNA (and other cellular contents) to be combined. The fused cells can then be induced to regenerate into whole plants, which will have traits derived from both parents. Some novel Chinese cabbage varieties developed through protoplast fusion have been marketed. Protoplast fusion has also been used to transfer *male sterility* (lack of pollen viability) from radishes into other *Brassica* species (cabbage, cauliflower, broccoli, canola), where it is used in the production of F1 hybrid seeds.

Marker-assisted Breeding

Marker-assisted (or *molecular-assisted*) breeding provides a dramatic improvement in the efficiency with which breeders can select plants with desirable combinations of genes. A marker is a "genetic tag" that identifies a particular location within a plant's DNA sequences. Markers can be used in transferring a single gene into a new cultivar or in testing plants for the inheritance of many genes at once. Markers can be based upon either DNA or proteins. DNA markers identify locations where the sequences differ among varieties or breeding lines. These can be locations within genes or in the DNA between genes, so long as they are unique sequences and differ between the plants of interest. Differences of this type are called polymorphisms, and there are a variety of ways to detect and use these signposts within the chromosomes.

Protein markers are based on differences in the proteins that are synthesised from the genes. Enzymes are proteins that catalyse specific biochemical reactions and can therefore be assayed for their activity. Plants often contain multiple copies of the same enzyme that differ slightly in size or charge, making it possible to separate them from each other. These are known as *isozymes*, and their sizes often differ among different genotypes. In these cases, the genes coding for the specific isozyme can be identified by following the inheritance of the isozyme.

Abundant proteins that are not enzymes, such as the seed storage proteins of cereal grains, can also be separated and stained to reveal specific size patterns characteristic of different genotypes. Both DNA- and protein-based markers have been widely used in plant breeding, but DNA-based markers are expected to predominate in the future.

Greater numbers of DNA-based markers can be identified to cover all regions of an organism's DNA, and they are not based on the developmental stage of the plant, as many proteinbased markers are. DNA-based markers can be visualised from seeds or seedlings in rapid screening tests, performed by automated robotic systems in advanced applications, and plants lacking the desired trait can be discarded before moving to more expensive or lengthy greenhouse and field trials.

When transferring a specific desirable gene from a wild plant into a crop plant via wide crosses, the simultaneous transfer of undesired genes from the wild plant is often a problem. In repeated crosses to the cultivated type, the percentage of wild type genes in each generation would be 50% in the first generation, 25% in the second, 12.5% in the third, 6.25% in the fourth, etc. Even after many generations, wild-type genes located close to the desired gene on the chromosome may still be present (i.e., are closely *linked* to the desired gene).

Marker-assisted breeding is helpful to identify plants that have inherited the desirable gene together with as few of the undesired genes as possible. One can think of the desired gene as the "needle in the (genetic) haystack." Traditional methods require sorting through the entire haystack to find the needle, while markers provide a "metal detector" to easily find the specific part of the haystack where the needle is located.

An isozyme marker speeded the development of commercial tomato varieties containing the *Mi* gene. The *Mi* gene was originally discovered in a wild relative of tomato and was transferred into cultivated tomatoes by sexual crosses and embryo rescue. Progress was slowed by the difficulty of screening for nematode resistance, which required that each plant be planted in soil infested with the nematodes, grown to maturity, and then examined for infection.

Later, it was discovered that a gene encoding a particular isozyme from the wild parent was located on the chromosome immediately adjacent to the *Mi* gene. This isozyme marker had been transferred together with the *Mi* gene in the original sexual crosses. As these genes were very close neighbors in the tomato DNA, virtually all plants that inherited the *Mi* gene also inherited the isozyme marker.

Once this linkage was discovered, breeders could predict whether an individual plant had inherited the Mi gene simply by testing which plants expressed the associated isozyme marker. Tests for the isozyme marker can be conducted quickly on very small seedlings, so only plants carrying the linked *Mi* gene needed to be grown to maturity. By using markers to speed the job of identifying plants that contained the *Mi* gene, breeders could focus their efforts on identifying plants with improved horticultural traits that did not inherit any undesirable genes from the wild tomato.

DNA markers are useful to test plants for inheritance of many different genes simultaneously. A simple way to appreciate this powerful tool for selection of improved crop plants is to visualise the UPC barcodes found on most products in the supermarket. The band pattern of the UPC code uniquely identifies the product and can be linked to a database of product traits such as cost, supplier, inventory, etc.

In a similar manner, the pattern of DNA markers allows the breeder to identify both desirable and undesirable genes that the plant may contain. As the number of traits that are desired increases, the number of plants that must be screened to identify plants that have the rare superior combinations of genes increases even more rapidly. By providing quick and efficient tests for many different genes, DNA markers are being widely used in breeding crop varieties having optimal combinations of desirable genes. DNA markers have been used for transferring pest resistance genes to cultivated varieties, assisting selection of complex multigene traits, aiding evaluation of regionally and seasonally optimised varieties, allowing genetic purity testing, and enabling proprietary variety protection and patent enforcement.

Recombinant DNA Techniques

DNA Cloning

Plant traits are encoded in the DNA of their genes. In 1973 it became possible to identify and multiply specific DNA molecules, based on discoveries by researchers at the University of California and Stanford University. They developed methods for isolating specific segments of DNA molecules, inserting them into other DNA molecules, and using bacteria as "biological copy machines" to produce large quantities of these DNA molecules. These recombinant DNA techniques enable researchers to isolate genes from any organism, amplify them in large quantities, and study their characteristics and functions. These techniques rely on restriction enzymes (such as EcoRI) that can recognise specific DNA sequences and cut the DNA strands at those sites. The ends of the DNA are cut unevenly, leaving an overhanging strand on each end. These are known as "sticky ends" because they are able to match up with other DNA strands that have complementary base sequences. Thus, if the same enzyme cuts two different DNA molecules, the ends of the molecules can match up and then be rejoined by another enzyme, DNA ligase. In this way, specific segments of DNA molecules can be cut and rejoined in different combinations. It is this ability to recombine different DNA molecules in the laboratory that leads to the term recombinant DNA technology.

In practice, a desired gene is cut from the DNA molecule using restriction enzymes and is inserted into a vector molecule that is then introduced into bacterial cells. The vector DNA (or plasmid) is reproduced as the cells grow and multiply, making many copies of the inserted DNA. In addition, the plasmid incorporates a separate gene that confers resistance to an antibiotic only to those cells containing the plasmid. Thus, if the cells are grown in a medium containing the antibiotic, only those cells having an intact plasmid, and thus also incorporating the transferred, or cloned, gene, will be able to survive. This selectable marker is essential to identify only the cells that contain the plasmid vector and allow them to multiply.

The cells can then be harvested and the DNA can be extracted and purified to obtain large quantities of the desired gene. To illustrate the basic concept of recombinant DNA methods, think of DNA as the biological equivalent of videotape. The instructions organisms need to grow and function are encoded on linear pieces of DNA, just as images are encoded magnetically in the linear pieces of videotape. When the videotape is played, the encoded information is converted electronically into the scenes displayed on the TV screen. Similarly, cellular machinery reads the DNA and converts the genetic information into proteins, which then carry out the functions necessary for life. Because DNA is chemically similar among all organisms, the instructions on these cloned pieces of DNA can be readily exchanged and "understood" between organisms as dissimilar as yeast and tomatoes. Just as technicians can cut, copy, and splice sections of videotape to create new scenes, scientists can now copy and exchange genes among organisms to introduce new characteristics.

In theory, any gene from any organism is potentially transferable to other organisms by recombinant DNA techniques, although this has been achieved in practice only among relatively few organisms. The inserted gene may give the organism the ability to do something that it could not do without the gene, such as resist attack by pests or produce a vitamin. Recombinant DNA techniques have essentially made all genes available to plant breeders for crop improvement, not just those found in a given species or its close relatives.

The ability to isolate segments of DNA and to determine the sequence of the component bases in them has resulted in an explosion of biological information. Genome projects determine the DNA sequences for all of the genes in an organism. Information from the recently completed Human Genome Project is expected to yield breakthroughs in understanding of human diseases and new therapeutic treatments. The genomic sequence for the model plant *Arabidopsis thaliana* was completed in December 2000, and genome projects are in progress for a number of major crop plants. These projects are revealing a remarkable similarity among the genes coding for specific proteins not only across most plants, but also in some cases across plants, animals, and microorganisms. These results demonstrate a fundamental consistency and conservation in the basic machinery of life.

Transformation

Once genes have been cloned, they can be transferred into the same or a different organism, producing a transgenic individual. Because of the simple genetic makeup of bacteria, transferring cloned genes into them is relatively easy. Thus, the first applications of recombinant DNA technology were to introduce useful genes into bacteria.

The first transgenic organism patent was issued in 1981 for a bacterium engineered to break down hydrocarbons. Now, a majority of all cheese is made using chymosin, an enzyme purified from microbes expressing a cloned chymosin gene, rather than using rennet extracted from calf stomachs. Most of the insulin used in human therapy for diabetes is produced by expressing the human insulin gene in microbes. Many other pharmaceuticals are manufactured in fungi, insect, or mammalian cells engineered to produce specific proteins.

Transfer of cloned genes into plants (*transformation*) is somewhat more difficult than transfer into bacterial and animal cells, but it can be accomplished by several methods. The most commonly used transformation method employs the bacterium Agrobacterium tumefaciens to transfer the desired DNA into the plant. When it enters a plant through a wound, this bacterium naturally transfers part of its DNA into the plant's chromosomes.

These bacterial genes cause the plant to make a structure to house the bacterium (a gall) and produce unique compounds that the bacterium consumes. In the 1980s, scientists discovered how to use *Agrobacterium* to transfer cloned genes into plant chromosomes while preventing the formation of galls. Desired genes can be spliced into the bacterial DNA, and then *Agrobacterium*, like a video editor, transfers them into the plant's chromosomes.

A second approach for production of transgenic plants, known as biolistics, coats the desired DNA onto microscopic metal beads and then fires these at high speed into pieces of plant tissue. The beads deliver the DNA into the plant cells, some of which will incorporate the DNA into their chromosomes. The transgenic cells resulting from these processes can be selected and regenerated into whole plants by tissue culture. These plants will then express the characteristics encoded by the cloned genes. Transgenic plant methods can be used to increase expression of an existing characteristic or to decrease expression of an undesirable characteristic in a crop plant.

As in the DNA cloning process, a critical part of the transformation process is the selection of transgenic individuals, as only a small fraction of cells or seed is likely to have incorporated the desired transgene. This is generally done by incorporating a gene conferring resistance to an antibiotic and then selecting only plant cells that survive growth on the antibiotic.

Some people have raised concerns about whether the widespread use of such genes will increase the development of antibiotic resistance in human pathogens and make current medicines obsolete. However, the likelihood of this is extremely remote. The antibiotics used in plant transformation were chosen because they are seldom used in human or veterinary medicine. Nonetheless, alternative types of marker genes have been developed and are replacing the antibiotic resistance method.

Transgenic plants are sometimes referred to as genetically modified organisms (*GMOs*). Despite the widespread use of this term, it is misleading because all of domesticated plants and animals have been genetically modified using a variety of traditional and modern technologies, as described above. These plants can be described more precisely as "transgenic plants" or plants containing cloned genes.

Recombinant DNA and transformation techniques allow plant breeders to use genes from essentially any source as tools for crop improvement. To enable rice grains to accumulate beta-carotene (which is converted into vitamin A when consumed by animals) and create the so-called "GoldenRice," scientists used genes from daffodil, pea, a bacterium, and a virus. Transgenic plant methods enable these four wellcharacterised genes to be inserted into a transgenic plant, producing a highly specific change in only the trait of interest. In contrast, many unknown genes are introduced when a breeder uses wide crosses to transfer a desired gene from a wild plant into a crop plant.

When the *Mi* gene for nematode resistance was transferred into tomato via wide crosses, the total amount of DNA exchanged was over 70 times larger than the *Mi* gene alone. Using the cloned *Mi* gene, only the nematode resistance trait would be introduced into other plants without transferring any additional DNA from the wild species. Woody tree and vine crops are difficult to improve by traditional breeding technology because it takes a number of years for a seedling to begin flowering, and the unique traits of specific varieties are hard to regain after sexual crosses.

Some wine grape varieties have been propagated for hundreds of years by cuttings, since plants produced from seed will segregate for genetic traits, often losing the distinctive characteristics desired in the variety. On the other hand, if only one or a few genes could be transferred into a traditional variety to confer resistance to a disease or insect, the valuable genetic combination represented by the variety would be preserved. Thus, transgenic methods may be valuable tools for improvement of these woody crops that constitute a significant fraction of California's agricultural abundance.

Plant breeders have not exhausted the genetic resources available for crop improvement through traditional techniques, and recombinant DNA methods have now greatly expanded those potential resources. Transgenic crop varieties already on the market are providing value to farmers and consumers while reducing the use of agrochemicals and enabling production methods that reduce soil erosion.

2

Fundamentals of Plant Breeding

By a process of selection, plant characteristics can be changed after many generations. There are two types of selection—natural and artificial. Natural selection is the process that occurs in nature whereby strong and well-adapted plants survive while weak and poorly adapted plants eventually die out. This process has taken place since the beginning of life on earth and it is still occurring in nature. Artificial selection is the process that humans use to obtain more desirable types of plants.

Thousands of years ago people learned that saving seed from the kind of plant they wanted to continue growing would increase the chances of getting a plant similar to the original. But our ancestors didn't know what their chances of success were nor did they understand the processes by which traits were changed or maintained.

It wasn't until the eighteenth and nineteenth centuries that humans began to understand the laws of heredity and the processes of plant reproduction. Even today these fundamentals aren't completely understood. But enough is known so that we can select plants for breeding with considerably more assurance of success than our primitive ancestors did.

Plant Reproduction

Plants reproduce in two ways—asexually and sexually. Asexual, or vegetative, reproduction occurs without the fusion of germ (reproductive) cells. In garden plants, asexual reproduction occurs when a part of the plant is separated from the parent plant and develops into a complete plant, as when strawberries produce runners which take root and form new strawberry plants. Asexual reproduction can be brought about artificially by means of leaf cuttings, stem cuttings, root cuttings, etc. Plants that originate from asexual reproduction are usually identical to the parent plant.

Sexual reproduction involves the union of a male and a female germ cell. From this union a seed—and ultimately a new plant—is produced. Sexual reproduction is the most common type of reproduction for garden plants. The plants originating from sexual reproduction are often quite different from their parents and from each other. Because of this possibility for variation, sexual reproduction of plants is the method used by plant breeders in developing new strains and varieties.

Parts of the Flower

Sexual reproduction of plants is carried on in the flower. The reproductive organs of the flower are the stamen and pistil (Fig. 1). A part of the stamen called the anther produces pollen grains, which contain male germ cells. At the base of the pistil is the ovary; it produces ovules, which contain the female germ cells. When the ovules are fertilized by the male germ cells, they develop into seeds in the ovary.

The petals, which add greatly to the beauty of flowers, serve such useful functions as aiding in pollination and protecting the sex organs from physical damage. The sepals, which often are green, cuplike structures, support the petals and protect the floral parts.

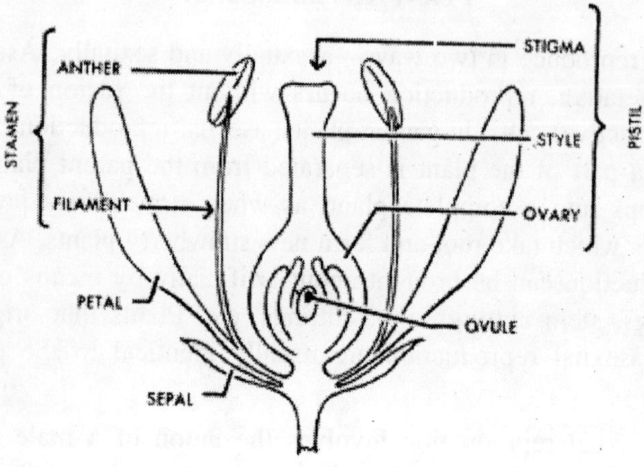

Figure 1. Parts of a flower

The flower parts vary greatly in size, shape, colour, and number, but generally it is easy to distinguish the pistil from the stamens. You can identify the stamens by the yellow powder (pollen) on their tips and the pistil by a bulge (the ovary) at its base. On an unusual flower, however, it might require careful study to tell them apart.

Types of Flowers

Usually a plant's reproductive organs are arranged in one of three ways—as a perfect flower, as an imperfect flower, or as a composite flower. A perfect flower contains both stamens and a pistil (Fig. 1) It is the most common type of flower. Some common perfect flowers are those of tomatoes, morning-glories, snapdragons, petunias, lilies, and irises.

An imperfect flower (Fig. 2) contains only one type of sex organ. Those containing the pistil are called pistillate flowers; those containing stamens are called staminate flowers. Both pistillate and staminate flowers may grow on the same plant, or the staminate flower may be on one plant and the pistillate flower on another.

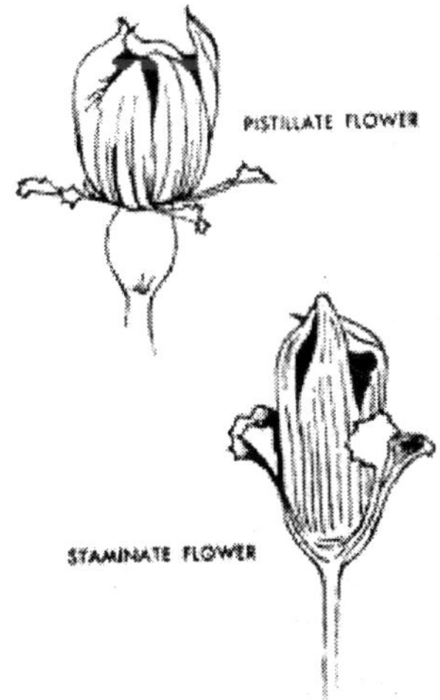

Figure 2. The flowers of squash are imperfect, each flower having only one type of sex organ.

Plants with pistillate and staminate flowers on the same plant are squash, corn, clematis, bittersweet, and begonias. Plants with staminate and pistillate flowers on different plants include asparagus, spinach, willow, and cottonwood. A composite flower (Fig. 3) is actually a cluster of small flowers, called florets, joined together in what is called a flower head.

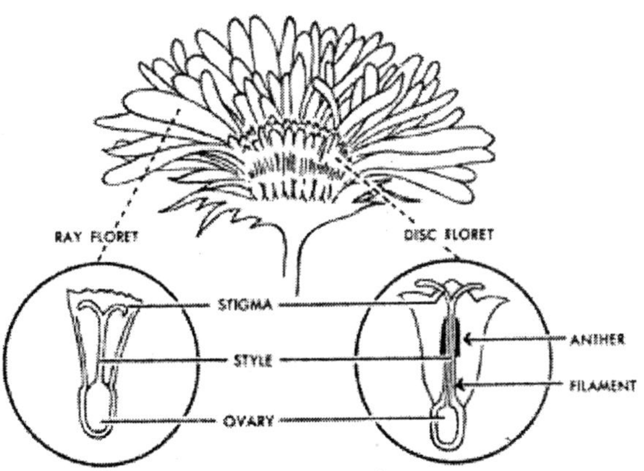

Figure 3. Cross section of a chrysanthemum.

The florets often resemble petals. Some florets have both stamens and pistils; they are called disc florets. Some have only pistils; they are called ray florets. A composite flower can be made up of only disc florets or of a combination of disc and ray florets. Typical composite flowers are zinnias, chrysanthemums, asters, and marigolds.

Pollination and Fertilization

Pollination is the transfer of pollen from an anther to a stigma. When the anther is mature it splits open and discharges the pollen. The pollen is then carried to the stigma by various natural means, the most common being wind and insects. In plant breeding, pollination is carefully controlled by humans.

There are two kinds of pollen transfer — cross-pollination and self- pollination. In cross-pollination the pollen is transferred from the anther of a flower of one plant to the stigma of a flower of another plant. In self-pollination the pollen is transferred from the anther to the stigma of the same flower or to another flower of the same plant.

Fertilization is the uniting of two germ cells. After the pollen has come to rest on the stigma, under favorable conditions the pollen develops a tube which grows downward through the style and into the ovule (Fig. 4).

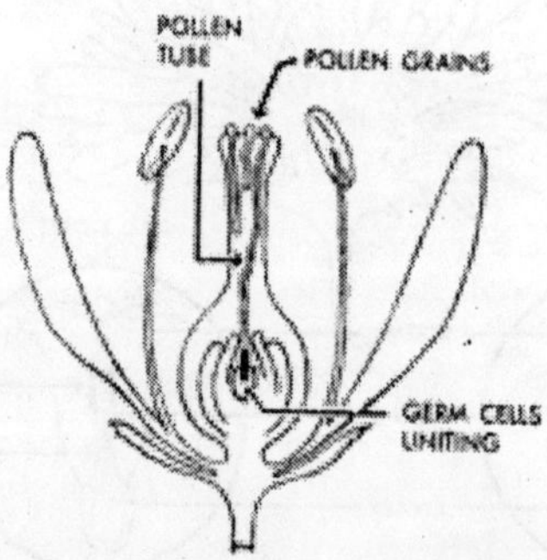

Figure 4. Fertilization takes place when the male and female cells unite in the ovule.

It is in the ovule that the male and female germ cells unite. Once fertilization has occurred development of the ovule begins. The result is a seed. The parent flower that furnishes the pollen is called the pollen parent; the one that bears the seed is called the seed parent.

General Breeding Techniques

The equipment required for plant breeding is relatively inexpensive and easy to use. Prepollination steps generally should begin just before the flower opens. If you wait until after it opens, it may be pollinated by natural means, which would make the flower useless for experimentation.

Since most plants bloom over a period of time, even if some of the flowers on a plant have bloomed before you have prepared them for breeding, there will probably be others that haven't opened yet and that you can still experiment with. Extremely high temperatures or moist conditions are harmful to pollen. For best results you should pollinate plants on dry days during the cool hours of the morning or as soon as the anthers have split open.

Selecting the Parents

Some plants have natural barriers to cross- or self-pollination. It is advisable to check for this before you begin breeding, for although you frequently can overcome these barriers, some plants cannot be artificially pollinated. An example of a barrier that can be overcome is the natural cross-pollination prohibitor of snapdragons; snapdragon flowers are constructed so as to prevent entry of wind-borne pollen from other flowers.

It is possible, however, to open the flower by hand. An example of a barrier that cannot be overcome is the self-pollination prohibitor of some orchids; the stigmas of certain orchids produce a substance which kills the pollen of flowers of the same plant. The mechanism that performs this cannot be removed without destroying the pistil.

The plants select for breeding should be sturdy and healthy. It is usually easier to tell which ones are healthy after a few flowers on the plant have bloomed. In choosing a pollen parent, select one that has a heavy yellow powder on the anther. This powder is the pollen. If you brush your fingernail against the anther, a trace of pollen should adhere to your nail. A fresh flower is more likely to have healthier pollen than one that has started to wilt or dry out.

In choosing a seed parent, examine the stigma. It should have either a glistening substance cn it that is sticky to the touch or a "hairy" surface. It is this substance or surface that retains the pollen, thus making fertilization possible. Once you have selected the pollen and seed parents, you are ready to begin pollination.

Prepollination Steps

The first step is to mark those flowers that are to serve as pollen parents and those that are to serve as seed parents. This can be done with coloured thread, one colour for the male and another colour for the female. Or can use paper labels, covered with varnish to protect them from the weather. Some plant breeders use bands designed for marking chickens.

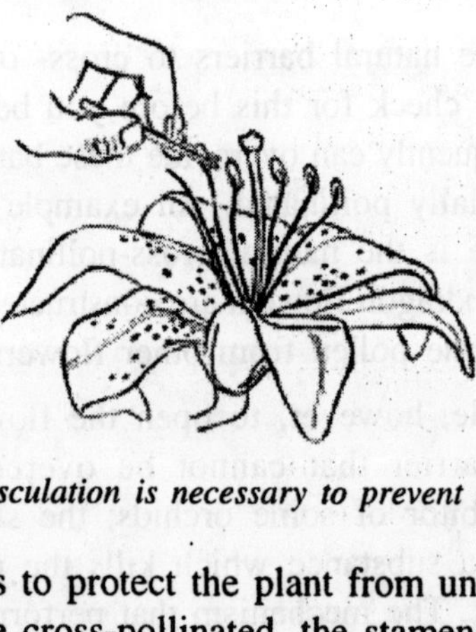

Figure 5. Emasculation is necessary to prevent self-pollination.

Your next step is to protect the plant from unwanted pollen. If the plant is to be cross-pollinated, the stamens will have to be removed to prevent the possibility of selfing. The removal of the stamen is called emasculation. It should be performed before the anthers split open to release pollen. This may require opening the flower by hand before it is ready to bloom. Emasculation can be accomplished by: (1) pinching off the stamens or anthers with tweezers (Fig. 5), or (2) snipping off the stamens or anthers with sharp-pointed scissors, or (3) removing the petals to which the stamens are sometimes attached. A magnifying glass will be particularly useful in emasculation.

Both the seed and pollen parents should be protected from contamination by foreign pollen. This can be done by one of the following methods:

Closing the flower: In many flowers, such as morning-glories, petunias, and lilies, the petals can be closed around the floral organs with a piece of soft wire, string, or rubber band (Fig. 6) Care should be taken not to tear the petals.

Covering the flower: Some flowers, such as composite flowers, cannot be closed. To protect them from unwanted pollen, you can cover the flower with a paper bag. Or, if you wish to observe the flower at all times, you can cover it with a cellophane bag. The bag should be held securely in place with a paper clip or string (Fig. 6).

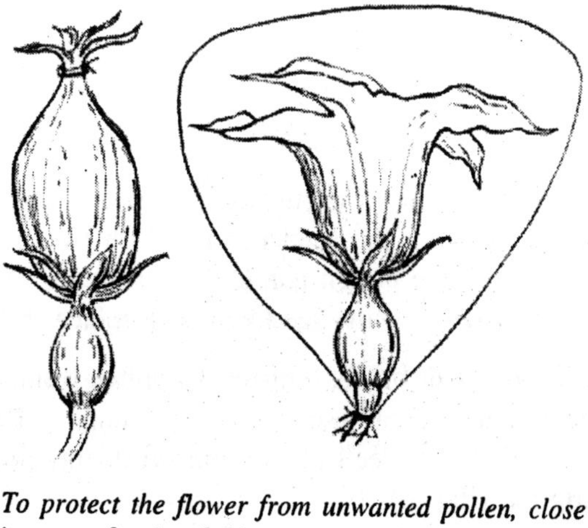

Figure 6. To protect the flower from unwanted pollen, close it with a string or soft wire (left) or cover it with a bag (right).

Pollination Steps

Crossing: There are several methods that can be used for crosspollinating flowers. Here are four of the most common methods:

1. Remove the stamens from the pollen parent with tweezers. Place the stamens in a small container. Remove the protector from the seed parent. Holding a stamen with the tweezers, gently brush the anther across the stigma. (Fig. 7). Replace the protector.

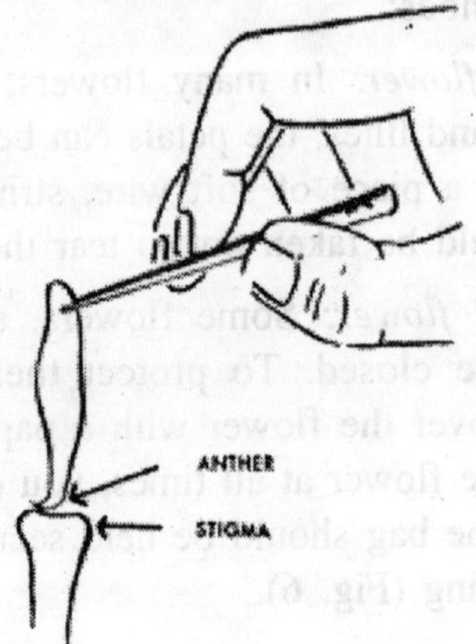

Figure 7. One way to cross-pollinate is to remove the stamens from the pollen parent and then, holding a stamen with tweezers, brush the anther across the stigma of the seed parent.

2. Cut the flower that is serving as the pollen parent. Remove the protector from the seed parent. With tweezers remove a stamen from the pollen parent and brush an anther gently across the stigma of the seed parent. Replace the protector.

3. With a camel-hair brush, transfer the pollen from the anthers of the pollen parent into a small container. Remove the protector from the seed parent and brush the pollen across the stigma. (Fig. 8)

4. Shake the bagged pollen parent so that the pollen is collected in the bag that is covering it for protection. Remove the bag from the pollen parent, being careful not to spill the pollen. Remove the protector from the seed parent and place the bag containing the pollen over the seed parent and shake the bag so that pollen falls on the stigmatic surfaces. This is usually done on corn.

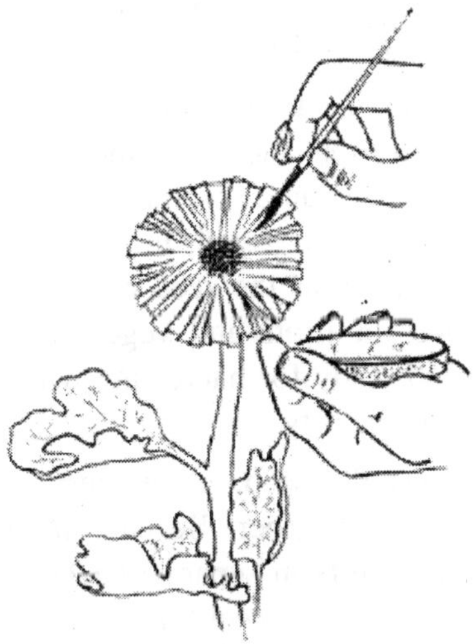

Figure 8. If the stamens are too small or too difficult to grasp, you can cross-pollinate by transferring the pollen from the anthers with a brush into a container and then brushing the pollen onto the stigma.

Each time you use different pollen, be sure to first wash with alcohol the camel-hair brush, tweezers, and any other item which might have touched some pollen. This step is very important to prevent pollination of the seed parent with unwanted pollen that has adhered to the equipment. After you wash the instruments be sure that they are dry before using them again.

Selfing: Procedures for self-pollinating flowers will depend on the type of flower. For perfect flowers, your job is done once you have closed or covered the flower, although you can sometimes help the pollen land on the stigma by shaking the flower once a day for several days after the pollen develops. Only those composite flowers containing both disc and ray florets can be self-pollinated. Since they have both pistils and stamens, they can be selfed in the same way as perfect flowers.

With imperfect flowers, will be able to self only those flowers that are on the same plant. In selfing imperfect flowers, the pollen from the staminate flower must be transferred to the stigma of the pistillate flower on the same plant.

Once pollination has been completed, a period must elapse during which fertilization and subsequent development of the fertilized ovule take place. When the seed is completely developed, it should be harvested. In plants grown for flowers, harvest the seed as soon as the seed pod is dry or when it just starts to split open. In fruits and vegetables the seed will be fully developed and ready for harvesting when the seed-bearing parts have reached maturity.

As you harvest the seed, place them in packets bearing the assigned offspring number. Be careful to keep seed from different crosses and selfs in separate packets. Air-dry the seed in a fairly warm spot for a week or so and then store them in a cool, dry place. The offspring can then be used for further study. Any seeds that are not planted should be saved in case of crop failure.

An important point to remember is that if both parents are pure, segregation usually will not become apparent until the F_2 generation, and you may have to breed several generations before you can obtain a plant that will breed true.

Breeding Experiments

The following crops have been selected to serve as examples of the various types of flowers encountered in plant breeding — perfect, imperfect, and composite. The techniques suggested here, with slight modifications, can be used for crossing other plants having similar flowering systems.

Corn

Corn is a good plant for the amateur plant breeder to experiment with because it is easy to work with and often shows visible changes in the kernels of the F_1 generation. Some of the varieties that can work with are yellow field corn, white field corn, sweet corn, popcorn, calico corn, and strawberry corn.

Corn has an imperfect flower, with both staminate and pistillate flowers on the same plant (Fig. 9). The staminate flower is the tassel; the pistillate flower is the ear. The silk, which has a hairy surface throughout most of its length that is receptive to pollen, is the corn's stigma.

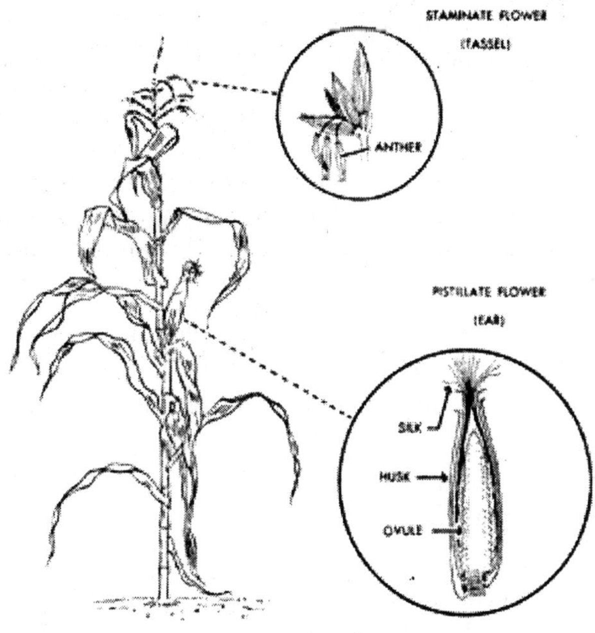

Figure 9. Corn

1. Decide on the varieties of corn that you wish to cross and plant the seeds in your experimental area. You can work with as many varieties as you wish, but if this is your first

experiment it will be best to limit yourself to three or four. Corn tassels lose their pollen in a relatively short time, so in order to have the pollen of the different varieties ready at about the same time, plant a few seeds of each variety weekly over a period of weeks.

2. As soon as the plants develop ear shoots and before the silks emerge, cover the shoots with a loosely fitted bag secured at the bottom with a paper clip or string. A clear plastic or glassine bag will allow you to watch the silks develop. Cover several ear shoots, for the more that are pollinated the greater your chances of getting desired results.
3. Keep a close watch on the ears. When the silks are visible the ear is ready for pollination. The appropriate tassel should then be bagged for use the next morning. The bag should be secured tightly at the base of the tassel to keep the pollen from falling out and to keep it from becoming contaminated with other pollen.
4. The following morning shake the bagged tassel vigorously to loosen the pollen. Remove the bag from the tassel, bending the tassel downward to prevent pollen from spilling out.
5. Remove the bag covering the silk. Place the bag from the tassel over the ear, being careful not to spill any pollen. Secure the bag to the stalk below the ear and shake.
6. Label each of the ears that have been pollinated with the name or number of the plants that are serving as seed and pollen parents and enter the cross.
7. Leave the bagged ear on the plant until the leaves dry. After the pollinated ear has dried, record any changes that may have occurred in the kernels. Then harvest the kernels and save them for future planting.

Squash

The flowers of squash are large and easy to work with. The

flowers are imperfect, with both the male and female flowers on the same plant. The pistillate flower is enlarged at the base while the staminate flower is borne on a long stem (Fig. 2).

Although some varieties of squash are cross-sterile, that is, they cannot be crossed, many others can be crossed. It is also possible to cross some varieties of pumpkin and squash. Remember that if you are crossing purë varieties, segregation won't occur until the F_2 generation; that is, the various colours and shapes that the offspring have inherited will not all show up until the offspring are self-pollinated and the seeds that they produce bear fruit.

1. Plant the different varieties of squash. Label the rows so that you will know which plants are to be seed parents and which are to be pollen parents. You won't have any difficulty in having the male and female parents ready for pollination at the same time, for squash produces flowers over a relatively long period.

2. The evening before the flowers open, place a bag over them or close the flowers with rubber bands or string (Fig. 6). It is easy to tell in the evening which flowers will open the following morning. The flowers will be slightly open, revealing the inner colour, which is brighter than the colour on the outside of the petals. Both pistillate and staminate flowers should be protected, to prevent bees from entering.

3. Early in the morning on the day that the flowers open, remove the stamens from the staminate flower and gently rub the anthers across the stigma of the pistillate flower. (Fig. 10.)

4. Replace the bag on the pistillate flower.

5. Label the pollinated pistillate flower and, enter the cross.

6. You can remove the protective covering when the petals have dried on the pistillate flower or when the ovary bulges through the bag.

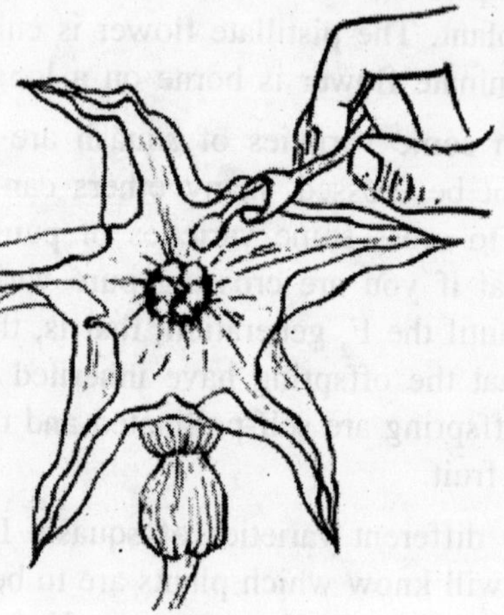

Figure 10. To pollinate squash, gently rub the anther across the stigma of the pistillate flower.

7. Harvest the seeds when the fruit is mature. Maturity is usually reached about 45 days after pollination. Save the seeds until the next year, when you will (1) plant them, (2) self-pollinate the resulting flowers, and (3) again harvest the seeds. The year after that, when you plant the seeds resulting from the self-pollination of the F generation, segregation will probably occur.

Tomatoes

The tomato has a perfect flower, with male and female organs both within the same blossom (Fig. 11). The sex organs are easy to tell apart and therefore it is relatively easy to cross tomatoes. For this experiment, choose tomatoes with two different fruit colours, such as red and yellow. If the tomatoes are pure, variations in colour will show up in the F_2 generation. Along with colour variations, there will probably be changes in fruit shape and other plant characteristics.

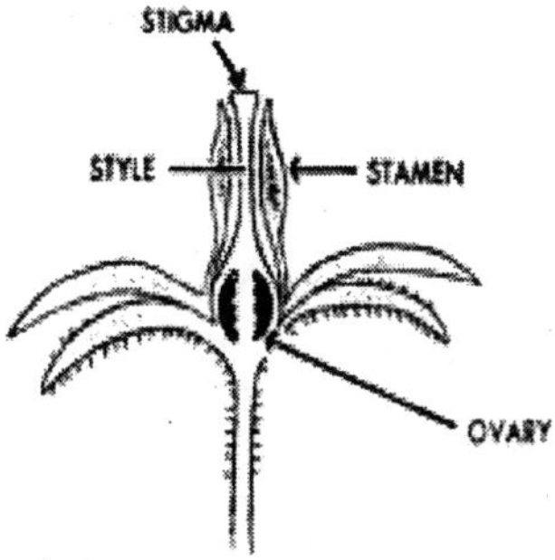

Figure 11. The reproductive parts of the tomato.

1. Plant the seed or young plants and label the rows so that you will know which will bear red fruit and which will bear yellow fruit. The tomato produces flowers for several weeks, so there will be no difficulty in having pollen and seed parents ready at the same time.
2. As soon as the flowers that will serve as seed parents begin to open, the plant should be emasculated. The stamens are fused to the petals, so all you need to do to emasculate the plant is to pull out the petals.
3. After you remove the stamens from the seed parents, cover the flowers securely.
4. The pollen parents that you wish to use should be in the same stage of opening as the seed parents. Cover the pollen parents at the same time that you cover the seed parents.
5. As soon as the pollen-parent flower is completely open, remove one of its stamens with scissors or tweezers. Remove the bag from the seed parent and gently brush the anther across the stigma.
6. Replace the bag that was removed from the seed-parent flower.
7. Label the flower that was pollinated (Fig. 12) and enter the cross.

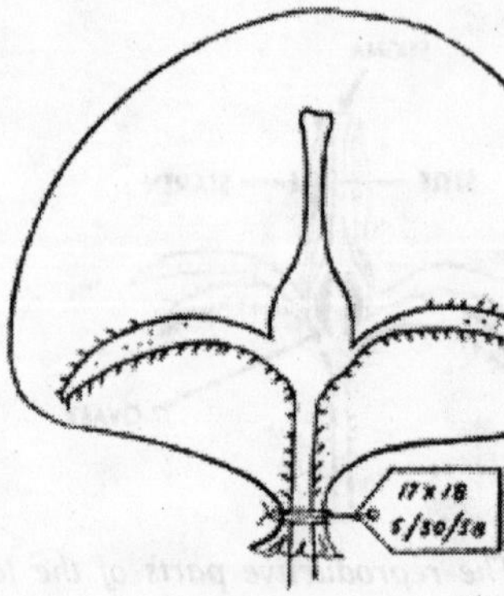

Figure 12. A tomato flower that has been emasculated, pollinated, and labeled.

8. After a week remove the bag from the seed parent and leave it off.
9. When the fruit is ripe, harvest and store the seeds for future planting and self-pollination.

Chrysanthemums

Chrysanthemums are composite flowers. If they are not available for crossing, you can use any other composite flower, such as zinnias or asters. Composite flowers are more difficult to cross than complete or incomplete flowers. If you are crossing plants for thfirst time, it would probably be best if you gained experience by first crossing a simpler flower, such as squash, iris, or lily, before trying to cross composite flowers.

1. Plant the seeds or young plants. Mums, zinnias, and asters bloom over a relatively long period, so there will be no difficulty in having pollen and seed parents ready at the same time.
2. Immediately after the seed-parent flowers open and before the pollen is mature, remove the disc florets with tweezers. As shown in Fig. 3, the disc florets are in the center of the

flower. They are easy to identify because they are often yellow in colour and they have both pistils and stamens, whereas the ray florets, which surround the disc florets, have only pistils. Remove some of the ray florets near the center of the flower just to be sure that you haven't missed any of the disc florets. If any disc florets are left on the flower, the flower is likely to be self-pollinated (Fig. 13).

Figure 13. When emasculating chrysanthemums, make sure that you remove all of the disc florets.

3. After removing the disc florets, cover the seed-parent flowers with bags.

4. As soon as the flowers of the pollen parents open, cover them with bags.

5. Pollination will be more effective if you trim the ray florets on the seed parents to just above the stigmas. This step can be done at any time after the disc florets are removed and before pollination takes placc (Fig. 14).

6. The flower is ready for pollination about four or five days after it opens. Shortly before pollinate the flower, dip a camel-hair brush in alcohol and allow it to dry.

7. Remove a flower from the pollen parent. With the camel-hair brush, gently brush the disc florets of the pollen-parent flower. The pollen should adhere to the brush. Then gently brush the trimmed ray florets of the seed-parent flower. Each time change pollen parents dip the brush in alcohol and let it dry.

Figure 14. Trim the ray florets for more effective pollination.

8. Replace the protective bags on the seed and pollen parents.
9. Label the pollinated flower and enter the cross.
10. Three or four days later, repeat step 7. Be sure to use the same pollen parent that you used the first time. Replace the bag on the seed parent.
11. Remove the bag from the seed parent about a week after the final pollination.
12. After the seed head has dried, harvest and store the seed.

3

GENETIC ENGINEERING OF PLANTS

The biotechnology refers to a broad spectrum of technologies including conventional plant selection and breeding—in which humans intervene in biological processes of genetic alteration and improvement. The products of genetic engineering are often called genetically modified organisms, or GMOs. All these terms refer to methods of recombinant DNA technology by which biologists splice genes from one or more species into the DNA of crop plants to transfer chosen genetic traits.

Genes are segments of DNA that contain information that in part determines the structure and function of a living organism. Genetic engineers are molecular biologists who manipulate this information, typically by taking genes from one species—an animal, plant, bacterium, or virus—and inserting them into another species, such as an agricultural crop. With the advent of genetic engineering of plants around 1983, it appeared that this new biotechnology would benefit and even revolutionise agriculture. The transfer of desirable genetic traits across species barriers has shown promise for solving problems in the management of agricultural crops Potential benefits include reduced toxic pesticide use, improved weed control resulting in less tillage and soil erosion, and water conservation.

GENE TRANSFER

Knowing how gene transfer occurs in the laboratory is key to understanding how the desired traits express themselves in the resulting crop plants, and why unintended effects, such as the production of allergens, may also be present or develop with subsequent generations of plants. Plants have elaborate defense mechanisms for dealing with foreign substances, including foreign DNA. To overcome these defense mechanisms, genetic engineers include pathological organisms in the complex DNA packages they create to carry the chosen gene into the host cell.

Generally, there are four parts to a DNA package:

1. Genes for the desired trait the payload. An example of such a trait is crop resistance to a given herbicide.
2. Genes for carrying the package into the host plant's DNA. This genetic carrier is called the vector and is usually taken from a bacterium that causes tumors in plants, Agrobacterium tumefaciens. Viruses are also sometimes used as gene carriers. These bacterial or viral vectors infect the new host cells, delivering the engineered gene into the DNA of the host plant.
3. Genes for ensuring that the genetic package will express the desired trait persistently. These genes turn on the desired trait and are called promoters. They are usually derived from the cauliflower mosaic virus (CaMV).
4. Genes for helping the biologist find the DNA segment in which the insertion has been successful. These genes are called markers and are resistant to antibiotics. When the antibiotic is applied to the new host's cells, the cells that survive are the ones carrying the successfully inserted antibiotic-resistant gene indicating a likelihood that the gene carrying the desired trait has been successfully inserted as well. Marker genes are usually derived from bacteria.

An alternative vector method uses a particle gun to shoot tiny metal particles with the DNA package attached into the new host cell. This eliminates the need for a bacterial vector, but adds to the potential for genetic disruption in the host cell.

The current methods of gene transfer are not precise. While scientists can control with relative exactness the trait gene that is inserted into a host plant genome, they cannot yet control its location, nor the number of copies that get inserted. Location of genetic material is important because it controls the expression of biological traits just as genes themselves do. Also, inserted DNA frequently contains multiple copies, which are often scrambled.

Multiple copies can "silence" genes, resulting in instability in the introduced genes. Because scientists can control neither the location nor copies of the genetic material they are inserting, they cannot entirely control the traits the host plant will express, nor can they guarantee genetic stability in subsequent generations. This can lead to unpredictable and undesirable effects, examples of which include plant infertility, the production of toxins, and reductions in yield and plant fitness.

Another characteristic that can result in unintended effects is called pleiotropy. Pleiotropy means that one gene controls multiple traits in an organism. In other words, the gene for a desired trait may also control other traits in the crop plant. Genetic engineers have relied on a simpler model of gene expression, assuming one gene controls one particular trait. But pleiotropy is common, and the interactions of genes with each other and their environment add further complexity. In short, it is difficult or impossible to predict what the effects of new genetic combinations will be.

The current marker and promoter genes of choice also create new hazards. The antibiotic-resistant marker genes carry the potential to increase the number of bacteria that are resistant to antibiotics. The viral promoter genes could combine with other infecting viruses to create new viruses.

The very powerful promoter commonly used, the cauliflower mosaic virus, can cause the inserted DNA package to be expressed out of proportion with the rest of the genetic code. With the particle gun method especially, this promoter can also jump out of the DNA package and land somewhere else in the host genome, causing disruption. The bacterial and viral vector genes can recombine to form active pathogens once again—either new ones, or old ones with renewed virulence, or with broader host specificity.

Each piece of the gene package described above carries with it the potential to disrupt non-target portions of the host plant's DNA, to create instability in the new genetic construct, or to result in unpredictable combinations that can create new substances, viruses, or bacteria. What this adds up to is the possibility of unintended effects, particularly in subsequent generations of the engineered plant.

Genetic engineering of crops is based upon the notion that genes alone determine species traits. But this assumption—and the resulting idea of simply transferring a single gene between organisms to achieve a single result—contradicts recent discoveries in genetic science. Inheritance and expression of traits are controlled by complex processes involving complicated genetic material that includes more than simple genes. For example, so-called junk DNA is now thought to play a significant role in determining an organism's traits.

Mapping of the human genome, completed in early 2001, revealed that humans contain about one-third the number of genes scientists expected to find. In fact, the 30,000 human genes they identified comprise only 1 percent of the total human DNA; they don't know what much of the other 99 percent is for. Only 300 genes in the human genome are not also in the mouse.

Commercial Transgenic Crops

While increased yields and improved nutritional value are among the promised benefits of transgenic crops, most genetically engineered crops now planted worldwide are designed either to:

1) survive exposure to certain herbicides or
2) kill certain insect pests.

Genetically engineered herbicide-tolerant crops have been altered to withstand being sprayed with broad-spectrum herbicides, with the idea that one application will take care of most types of weeds without killing the crop. Insecticidal crops contain genes of the soil bacterium Bacillus thuringiensis (Bt). These Bt genes cause the plants to produce a chemical toxic to the European corn borer, the cotton bollworm, and other caterpillars.

Herbicide-tolerant crops accounted for 71 percent of the acreage planted, worldwide, to genetically engineered crops in 1998 and 1999. Pesticidal crops, or a combination of pesticidal and herbicide-tolerant crops, accounted for most of the remaining acreage. More than 100 million acres of the world's farmland were planted with transgenic crops in the year 2000.

The United States, Argentina, and Canada are the world's leading producers of genetically engineered crops, with the U.S. first at 71 percent of the world's total acreage. Transgenic crops are being shipped to or experimented with in many other countries, including China, India, Australia, and South Africa.

In 1999, one-quarter of U.S. farmland was planted to transgenic crops. To date, soybeans and corn cover the most transgenic acres. In 2001, more than 60 percent of soybeans planted in the U. S. were estimated to be transgenic specifically, tolerant to Monsanto's broad-spectrum herbicide. Transgenic corn planted in the U.S. offers both insecticidal and herbicide-tolerant traits, though Bt insecticidal corn predominates.

Other large-acreage transgenic crops include cotton and canola. Most transgenic cotton is herbicide-tolerant, though some varieties have the Bt trait; transgenic canola is herbicide-tolerant. The first transgenic wheat, planned for commercial introduction in 2003, is Roundup-tolerant.

Other traits engineered into commercial transgenic varieties include disease resistance, high pH tolerance, and several nutritional, taste, texture, and shelf-life characteristics. Other transgenic crops currently on the U.S. market include tomatoes, potatoes, sunflowers, peanuts, and sweet mini-peppers. More transgenic crops, including rice, are under development for commercial release in the next several years. However, there is evidence that biotechnology firms may be changing course because of deepening consumer skepticism and tighter regulation worldwide, which are increasing the costs and business risks of new transgenic crop introductions.

Ecological Issues

Gene flow to neighbouring crops and to related wild species

Ecological scientists have little doubt that gene flow from transgenic fields into conventional crops and related wild plants will occur. Gene flow from transgenic to conventional crops is of concern to farmers because of its potential to cause herbicide resistance in related conventional crops. In western Canada, three different herbicide-resistant canola varieties have cross-pollinated to create canola plants that are resistant to all three types of herbicide. Through what is called gene-stacking, this new triple resistance has turned volunteer canola into a significant weed problem.

Gene flow from transgenic crops to wild relatives creates a potential for wild plants or weeds to acquire traits that improve their fitness, turning them into super weeds. If jointed goatgrass—a weedy relative of wheat acquires the herbicide-tolerant trait of Roundup Ready wheat, it will thrive in crop fields unless applications of other herbicides are made. There is already evidence of such outcrossing from herbicide-resistant wheat to jointed goatgrass. Frank Young and his colleagues at Washington State University found that imidazolone-resistant wheat outcrossed to goatgrass in one season. Other traits that wild plants could acquire from transgenic plants that would increase their weediness are insect and virus resistance.

Because of their experience with classically bred plants, few scientists doubt that genes will move from crops into the wild: seven of the world's thirteen most important crop weeds have been made weedier by genes acquired from classically bred crops. Because gene flow has the potential to affect farmers' crop and pest management, crop marketability, and liability, more research needs to be done to determine the conditions under which gene flow from transgenic plants is likely to be significant.

Pesticide resistance

Bt has been widely used as a microbial spray because it is toxic only to caterpillars. In fact, it is a pest management tool that organic farmers depend on one of the few insecticides acceptable under organic rules. Unlike the commercial insecticide spray, the Bt engineered into crop plants is reproduced in all, or nearly all, the cells of every plant, not just applied on the plant surface for a temporary toxic effect. As a result, the possibility that transgenic Bt crops will accelerate insect pests development of resistance to Bt is a serious concern.

Antibiotic resistance

How gene transfer is accomplished, the use of antibiotic-resistant marker genes for the delivery of a gene package into a recipient plant carries the danger of spreading antibiotic-resistant bacteria. The likely result will be human and animal health diseases resistant to treatment with available antibiotics. Research is needed on antibiotic resistance management in transgenic crops. Already the European Commission's new rules governing transgenic crops stipulate phasing out antibiotic-resistant marker genes by the end of 2004.

Effects on beneficial organisms

Evidence is increasing that transgenic crops either directly or through practices linked to their production are detrimental to beneficial organisms. New studies are finding that Bt crops exude Bt in concentrations high enough to be toxic to some beneficial soil organisms. In addition, a study by University of Arkansas scientists has shown that root development, nodulation and nitrogen fixation are impaired in some varieties of Roundup

Ready soybeans. The reason is that the beneficial rhizobium responsible for nitrogen fixation in soybeans is sensitive to Roundup. It also appears that disruption of beneficial soil organisms can interfere with plant uptake of phosphorus, an essential plant nutrient. Beneficial insects that prey on insect pests can be affected by insecticidal crops in two ways.

First, the Bt in transgenic insecticidal crops has been shown in some laboratory studies to be toxic to ladybird beetles, lacewings, and monarch butterflies.

Second, because the insecticidal properties of Bt crops function even in the absence of an economic threshold of pests, Bt crops potentially can reduce pest populations to the point that predator species are negatively affected.

Reduced crop genetic diversity

As fewer and larger firms dominate the rapidly merging seed and biotechnology market, transgenic crops may continue the trend toward simplification of cropping systems by reducing the number and type of crops planted. In addition, seed-saving, which promotes genetic diversity, is restricted for transgenic crops.

Food Safety

Because food safety is predominantly a concern of consumers. However, because food safety issues do impact marketing and international trade, here is a short list of consumer safety concerns about transgenic food:

— possibility of toxins in food

— possibility of new pathogens

— reduced nutritional value

— introduction of human allergens

— transfer of antibiotic resistance to humans

— unexpected immune-system and genetic effects from the introduction of novel compounds

It is in part because of these concerns that consumer demand for organically grown crops continues to increase.

Farm Management Issues

The most widely planted transgenic crops on the market today can simplify short-term pest management for farmers and ranchers. In the case of herbicide-tolerant crops, ideally farmers can use a single broad-spectrum herbicide for all their crop weeds, though they may need more than one application in a season. By planting insecticidal crops, farmers can eliminate the need to apply pesticides for caterpillar pests like the European corn borer or the cotton bollworm, though they still have to contend with other crop pests. While these crops offer handy pest control features, they may complicate other areas of farm management. Farmers who are growing both transgenic and conventional varieties of the same crop will need to segregate the two during all production, harvesting, storage, and transportation phases if they are selling into differentiated markets or plan to save their own seed from their conventional crops.

To minimise the risk of gene flow from transgenic to adjacent conventional crop fields, federal regulations require buffer strips of conventional varieties around transgenic fields. Different transgenic crops require different buffer widths. Because the buffer strips must be managed conventionally, producers have to be willing to maintain two different farming systems on their transgenic fields. Crops harvested from the buffer strips must be handled and marketed as though they are transgenic.

Planted refuges—where pest species can live outside fields of insecticidal and herbicide-tolerant transgenic crops—are also required to slow the development of weed and insect pest resistance to Bt and broad-spectrum herbicides. These refuges allow some individuals in the pest population to survive and carry on the traits of pesticide susceptibility. Requirements governing the size of refuges differ according to the type of transgenic crop grown.

Farmers growing herbicide-tolerant crops need to be aware that volunteer crop plants the following year will be herbicide-resistant. This herbicide resistance makes no-till or direct-seed systems difficult because volunteers can't be controlled with the same herbicide used on the rest of the crop. In a no-till system that relies on the same broad-spectrum herbicide that the volunteer plants are resistant to, these plants will contaminate the harvest of a following conventional variety of the same crop, a situation farmers need to avoid for two reasons.

First, the contamination means a following conventional crop will have to be sold on the transgenic market. This leads to the second reason. If farmers grow and market a transgenic crop for which they do not have a technology agreement and did not pay royalty fees, they face possible prosecution by the company that owns the transgenic variety.

Many farmers have been charged with theft of a company's patented seed as a result of contamination in the fields. Farmers growing insecticidal crops need to recognise that insect pressure is difficult to predict and may not warrant the planting of an insecticidal variety every year. In a year when pest pressure is low, the transgenic seed becomes expensive insurance against the threat of insect damage.

Farmers growing transgenic crops need to communicate with their neighbours to avoid contaminating their fields and to ensure that buffers are adequate. In Maine, farmers growing transgenic crops are now required by law to be listed with the state agriculture department, to help identify possible sources of cross-contamination when it occurs. The new law, L.D.1266, also requires manufacturers or seed dealers of genetically engineered plants, plant parts or seeds to provide written instructions to all growers on how to plant, grow, and harvest the crops to minimise potential cross-contamination of non-genetically engineered crops or wild plant populations.

Farm management issues common to all transgenic crops include yield, cost, price, profitability, management flexibility, sustainability, market acceptance, and liability.

Crop Yield, Costs, and Profitability

Some farmers will get higher yields with a particular transgenic crop variety than with their conventional varieties, and some will get lower yields. The same is true for costs. Some producers will see their overall profitability rise, and some will see it drop. If the results of research studies so far on yield, chemical use, costs, crop prices, and profitability of these crops seem contradictory, it is because farmers and farms are not all alike. Neither are the conventional varieties used for comparison.

Some yield, cost, and profitability trends do appear to be emerging from the growing body of research data for transgenic crops, however. Wallace Center report, Roundup Ready soybeans were designed simply to resist a particular chemical herbicide, not to increase yields. In contrast, Bt corn and cotton, by resisting insect pests, may result in higher yields from reduced pest pressure.

Herbicide-tolerant soybeans appear to suffer what's referred to as a "yield drag". Again, in some areas and on some farms this tendency of Roundup Ready soybean varieties to yield less than their comparable, conventional counterparts varies, but overall, they appear to average yields that are five to ten percent lower per acre. Impaired root development, nodulation, and nitrogen fixation likely account for this yield drag.

Drought conditions worsen the effects. The bacterium that facilitates nodulation and nitrogen fixation in the root zone apparently is sensitive to both Roundup and drought. In addition, University of Missouri scientists reported problems with germination of Roundup Ready soybeans in the 2001 crop year.

Yields of herbicide-tolerant cotton are reportedly not significantly different from those of conventional cotton. Herbicide-resistant transgenic canola varieties yield less on average than conventional canola varieties. Transgenic canola costs less than conventional canola to produce, but because of its higher yields conventional canola returns more profit per acre.

Insecticidal Bt corn and cotton generally yield higher—in most years for some regions—according to USDA Economic Research Service data from 1996 to 1998. Bt cotton, especially, outpaces yields of conventional cotton by as much as 9 to 26 percent in some cases, though not at all in others. Yield increases for Bt corn have not been as dramatic. Time will tell whether farmers can expect yield increases or decreases in the long run with these and other transgenic crop varieties.

Changes in Chemical Pesticide Use

One of the promises of biotechnology is that it will reduce pesticide use and thereby provide environmental benefits and reduce farmers' costs. The herbicide-tolerant and insecticidal varieties are designed specifically to meet these goals. Studies estimate a two to three percent decrease in U.S. pesticide use, but the effects vary widely by crop, region, and year. Increased future pesticide use resulting from the buildup of resistance to heavily used herbicides is a long-term concern.

Pesticide use depends on the crop and its specific traits; weather; severity of pest infestations; farm management; geographic location of the farm; and other variables. As a result, conclusions drawn by various studies analysing pesticide use on transgenic crops remain controversial. The crop for which studies are showing the largest decrease in pesticide use is Bt cotton, with Bt corn resulting in only small changes. Herbicide-tolerant cotton has also resulted in little change in herbicide use. American Soybean Association president Tony Anderson agrees that the developing resistance of weeds to herbicides such as Roundup is a problem.

The Wallace Center report emphasises the importance of ongoing monitoring of pesticide use data. If farmers abandon integrated pest management, which utilises a variety of pesticide and cultural control methods, in favour of the simplified control offered by herbicide-resistant and insecticidal transgenic crops, then early findings of reduced pesticide quantities and toxicity may not hold over the long run.

Farmers need to consider all the factors that determine profitability. No single factor can tell the whole story. Transgenic crop seeds tend to be more costly, and farmers have the added expense of a substantial per-acre fee charged by the owners of transgenic varieties. These costs have to be considered along with input cost changes whether herbicide or insecticide use and costs go down, go up, or stay the same.

Prices for some transgenic crops in some markets are lower than prices for comparable conventional crops, though rarely are they higher. Farmers need to watch the markets. Some buyers will pay a premium for a non-transgenic product, though as transgenic seeds find their way into conventional transportation, storage, and processing streams, these premiums may disappear along with confidence that GMO-free products are in fact truly free of engineered genes.

Issues in Marketing and Trade

Buyer acceptance is a significant marketing issue for farmers raising transgenic crops. Farmers need to know before they plant what their particular markets will or won't accept. Since most grain handlers cannot effectively segregate transgenic from non-transgenic crops in the same facility, many companies are channelling transgenic crops into particular warehouses. Farmers need to know which ones and how far away those are.

Many foreign markets tend to be more leery of transgenic products than domestic markets, though trade in transgenic livestock feed is more liberal than trade in transgenic human food. The widespread contamination of corn in the U.S. with the Cry9C Bt transgene, which is not approved for human consumption, has resulted in even further resistance on the part of buyers to purchasing transgenic products for human food.

Some countries have banned altogether the production and importation of transgenic crops and food products (Sri Lanka and Brazil), while many others have put in place partial bans or mechanisms to slow their approval for importation or production. While the European Union officially lifted its two-year moratorium on the introduction of new transgenic crops in February 2001, during the debate over labelling and traceability regulations the moratorium remains in effect. Under the proposed new EU requirements, all foods and animal feed derived from GMOs have to be labelled and, in the case of processed goods, records have to be kept throughout the production chain allowing the GMO to be traced back to the farm. If approved, the new regulations will hamper the export of U.S. farmers' products to the EU because the U.S. does not require traceability or labelling of transgenic crops.

While the U.S. does not require mandatory labelling of processed food containing transgenic ingredients, the European Union, Russia, Czech Republic, Japan, South Korea, Taiwan, Australia, New Zealand, and Ecuador do. Because many domestic merchandisers of agricultural commodities do not segregate transgenic from conventional crop varieties, it is impossible for them and the farmers that supply them to serve these food markets.

Twenty percent of corn and 8 percent of soybeans produced in the U.S. are exported, and more than 80 percent of these crops is used in animal feed. Overseas market restrictions on transgenic crops in the feed sector are less stringent than those for food. In contrast, 48 percent of U.S. wheat is exported, with nearly half of this food crop going to countries that are resistant to buying transgenic food, particularly Japan and European nations. Because wheat producers are so dependent on exports, they are approaching the introduction of the first transgenic wheat, expected as soon as 2003.

The Japanese milling industry has made it clear that it does not want transgenic products. As a result, Monsanto has promised not to introduce its Roundup Ready wheat until Japan gives its approval. At least two states, North Dakota and Montana, have considered legislation that would place a state moratorium on the introduction of transgenic wheat. For more about the wheat industry's reaction to the anticipated introduction of Roundup Ready wheat.

In addition to national and international policies on the use and importation of transgenic crops, many countries' processors and retailers have set their own corporate policies. Major retail chains in Europe and the U.S. have declared their commitment to avoiding the purchase of transgenic products, both feed and food.

For buyers who don't want transgenic products, the likelihood of contamination of non-transgenic products with engineered genes is troublesome. Since 1997, the European Union has effectively barred U.S. corn imports over the possibility that genetically engineered varieties unapproved in the EU have mixed with sanctioned crops. This has cost American farmers access to a $200 million-a-year market.

Organic farmers face even bigger marketing and trade risks, since their buyers can accept no transgenic contamination. The organic industry does have a system for segregation, but recent tests for transgenic material in organic products demonstrate that it is not immune to contamination from conventional systems. The markets organic farmers have enjoyed, and those that producers of non-GE conventional crops could build upon, will likely disappear if contamination of their products continues to spread.

Until the issues of trade in transgenic crops and food products, and their labelling and traceability, are resolved in international trade arenas, the markets for U.S. transgenic crops will remain uncertain. Changes in what any particular country will or won't accept are occurring all the time, and producers of crops for export are well advised to stay informed of these changes. Segregation of transgenic crops and liability for cross-contamination of conventional and organic crops are issues that obviously overlap with those of marketing and trade.

The need to separate transgenic crops from conventional and organic ones opens farmers to liability for their product at every step from seed to table. Effective systems for segregation do not exist at present, and will be costly to develop and put into place. Farmers may well end up bearing the added costs of crop segregation, traceability, and labelling. In the meantime, farmers who grow transgenic varieties and, ironically, those who do not are liable for transgenic seeds ending up where they aren't wanted: in their own non-transgenic crop fields, in neighbours' fields, in truckloads of grain arriving at the elevator, in processed food products on retail shelves, and in ships headed overseas.

Farmers who choose not to grow transgenic varieties risk finding transgenic plants in their fields anyway, as a result of cross-pollination via wind, insects, and birds, which may bring pollen from transgenic crops planted miles away. Besides pollen, sources of contamination include contaminated seed and seed brought in by passing trucks or wildlife.

Those farmers whose conventional or organic crops are contaminated, regardless of the route, risk lawsuits filed against them by the companies that own the proprietary rights to seed the farmer didn't buy. Likewise, farmers who grow transgenic crops risk being sued by neighbours and buyers whose non-transgenic crops become contaminated. Because contamination by transgenic material has become so prevalent in such a short time, all farmers in areas of transgenic crop production are at risk.

Insurance, the most common recourse for minimising potential losses because of liability, is not available to the nation's farmers for this risk because insurance companies do not have enough information to gauge the potential losses. Most of the farmers who have been accused by transgenic seed companies of illegally growing and harvesting their proprietary transgenic varieties have paid fines to the companies rather than go to court to defend themselves.

A Texas organic farm's corn, assumed to be GE-free, was purchased by a processor who made it into organic tortilla chips.

Only after the product had been sold and shipped to European retailers was it discovered to be contaminated with transgenic corn. The processor had to recall its product at a cost of over $150,000. The processor chose not to sue the organic farmer, but could have. The retailer, in turn, apparently did not sue the processor. Had it done so, the liability for the retailer's loss could have fallen on both the processor and the farmer.

Producers of transgenic crops need to take all possible precautions against spreading pollen and seed to their own and others non-transgenic fields and markets. A full risk assessment and legal clarification of the distribution of liability among farmers, seed companies, grain handlers, processors, and retailers is needed before farmers can rest assured that transgenic crops won't result in lawsuits against them. While transgenic crop varieties are generally the property of private corporations, those corporations often contract with public-sector agricultural research institutions for some of their development work. In fact, private investment in agricultural research, including germplasm development, has surpassed public investment in recent years.

Because intellectual property rights (patents) apply to living organisms, making them private property, the free flow of scientific information that has historically characterised public agricultural research is being inhibited. What are the implications for the future of agriculture and society of the secrecy that now surrounds so much of what was formerly shared public knowledge? The issues are of particular concern to farmers and consumers who would benefit from research into alternative technologies that are less costly (in every way), less risky, and more equitable. Equity requires that the economic benefits and risks of transgenic technology be fairly distributed among technology providers, farmers, merchants, and consumers.

For long-term sustainability, farmers need research that focuses on farms as systems, with internal elements whose relationships can be adjusted to achieve farm management goals. In contrast, transgenic crop research so far has focused on products that complement toxic chemical approaches to control of individual pest species. These are products that can be commercialised by large agribusiness or agri-chemical interests and that farmers must purchase every year. This research orientation only perpetuates the cost-price squeeze that continues to drive so many out of farming.

Regulation of Transgenic Crops

Much of the controversy over transgenic crops, both internationally and in the U.S., is in part a result of how the U.S. regulates transgenic crops. The federal government has determined that the commercial products of agricultural biotechnology are substantially equivalent to their conventional counterparts and that therefore no new regulatory process or structure is needed for their review and approval.

Currently, three federal agencies regulate the release of transgenic food crops in the U.S.: the U.S. Department of Agriculture's Animal and Plant Health Inspection Service (USDA-APHIS), the U.S. Environmental Protection Agency (EPA), and the U.S. Food and Drug Administration (FDA).

USDA-APHIS: The USDA looks at how a transgenic plant behaves in comparison with its unmodified counterpart. Is it as safe to grow? The data it uses are supplied largely by the companies seeking a permit for release of the new crop. Under fast-track approval, a process in place since 1997, companies introducing a crop similar to a previously approved version need give only 30 days advance notice prior to releasing it on the market.

EPA: The EPA regulates the pesticides produced by transgenic crops, such as the Bt in Bt corn and cotton. It does not regulate the transgenic crops themselves. In contrast to its regulation of conventional pesticides, the EPA has set no tolerance limits for the amount of Bt that transgenic corn, cotton, and potatoes may contain.

FDA: The FDA focuses on the human health risks of transgenic crops. However, its rules do not require mandatory pre-market safety testing or mandatory labelling of transgenic foods.

Initially, the U.S. regulatory process for transgenic food crops required product-by-product reviews. Now, however, to simplify and speed up the process, new products can be approved based on the experience gained in reviewing earlier products. The regulatory process, in fact, may not answer most questions about the environmental and human health risks of commercial production of these crops.

Central to the policy of substantial equivalence is the assumption that only the end product of transgenic technology is of concern not the process of genetic modification. Canada has adopted a similar approach. Europe and other U.S. trading partners, however, have taken a more conservative approach. They focus on the process of genetic modification the source of many of the environmental and human health risks of greatest concern

How these different approaches play out in reality can be summed up simply: The U.S. and Canada assume a product is safe until it is proven to carry significant risk; the European Union, which follows the precautionary principle, assumes the same product may carry significant risk until it can be proven safe. The science used by the two approaches is not fundamentally different. The difference is in the level of risk the different societies and political systems are willing to accept.

Implications for Sustainable Agriculture

In contrast to the ecological approach of sustainable agriculture, the current generation of transgenic crops follows a pest management model like that employed for chemical pesticides through interventions that are toxic to pests, the Wallace Center report points out. This single-tool approach is likely to fail in the long run because pests will successfully develop resistance that allows them to thrive.

Standard plant-breeding methods can potentially solve many of the same problems in agriculture that genetic engineers are working on, though there are areas in which genetic engineering can enhance traditional plant breeding. Armed with the map of an organism's genetic code, scientists can test which genes are in a plant to select more easily which ones to cross-breed.

4

DNA ANALYSES IN PLANT BREEDING

The possibilities of improving plants expanded only in the last century as a result of Mendel's investigations into hereditary traits in peas and subsequent discoveries of the genetic basis of inheritance. Among them chromosomal theory, explanation of mutations and also the discovery of DNA structure and function played an important role. Induced mutagenesis, i.e. changes in the genetic basis of the plant using chemical compounds or radioactivity, was employed after World War Two. So-called "Green Revolution" in the 50's involved the simultaneous development of new varieties of crop plants and altered agricultural practices that greatly increased crop yields.

Once the genetic basis of heredity was understood, plants with different desirable traits were systematically selected and crossed in order to produce new varieties that combined better characteristics of the donor material. The end-use quality of crops has also been improved with respect to e.g. protein or oil contents. The demands of the market have called for speeding up the breeding process and developing cultivars with high and stable yield, superior or altered qualities. Two methods and their combinations appeared at thc cnd of the last century—genetic transformation and marker assisted selection. Whereas the first approach offers a rapid method combining genetic materials from different species, e.g. transmission of a bacterial gene conferring herbicide tolerance into several plant species, the other intends to use the information on the structure and function of plant genome to efficiently assort parental material and speed up selection of the best progenies using molecular markers.

Molecular markers are molecules that could be used to trace a desired gene(s) in examined genotypes. In fact a piece of DNA or a protein can be used as a marker. Earlier approaches that made selection of specific traits easier were based on the evaluation of morphological traits, isozymes, storage proteins like glutenins, gliadins, hordeins, etc. However, DNA markers seem to be the best candidates for efficient evaluation and selection of plant material. Unlike protein markers, DNA markers segregate as single genes and they are not affected by the environment. DNA is easily extracted from plant materials and its analysis can be cost and labour effective.

Before molecular markers can be used, they have to be developed using DNA technologies, appropriate plant material and suitable algorithms. Selection of suitable plant population is as important as selection of suitable DNA technique(s). There are two basic categories of molecular markers:

(1) Markers segregating and determining the presence of a single, dominant or recessive, gene and

(2) QTL associated markers. It is much easier and cheaper to develop marker(s) for a single gene inherited trait than Quantitative Trait Loci (QTLs).

Generally, development of DNA markers consists of several steps. First it is necessary to analyse the nature of studied trait(s) e.g. by genetic analysis. After appropriate mapping the population has to be developed. A much larger population (e.g. a set of DH lines, advanced backcross population) is required for QTL mapping in comparison with a single gene mapping. The mapping population has to be evaluated in the field and/or in a laboratory. It means that each individual line must be tested for a selected trait value. The same lines are analysed by DNA techniques. Based on field/laboratory analysis and DNA tests putative marker(s) are identified: mapping is usually done by establishing a statistical association between molecular marker genotype and phenotype.

Most statistical approaches require a continuous distribution of the response variable for QTL development, but odds utilisation is also possible. Finally marker(s) have to be validated using additional plant material.

The breeding process consists of several steps. First, parental material must be carefully selected. Lines are usually chosen from the currently available gene pool of contemporary varieties. Besides that, in some cases wide relatives or exotic germplasm are used to introduce a new trait, mostly disease resistance. To test genetic resources for their productivity, quality parameters and stress tolerances field trials and chemical/physical/biochemical tests are employed. Field trials are usually time consuming, therefore molecular markers and DNA technology are used to assess diversity in the gene pool, to identify genes of interest and to develop a set of markers for the screening of progenies.

Parental lines that are used for crossing are always carefully selected. According to the breeding plan and breeding aim more or less related lines are used. There are several ways of estimating genetic similarity of cultivars. Basically they can be divided into morphologybased, pedigree-based and until now less frequently used marker-based methods. The value of some marker systems, such as isozymes, is questionable. DNA based technologies are more suitable. However, the estimated levels of polymorphism of the varieties widely varied with techniques used.

Much has been expected from DNA technology when hybrid breeding is considered. Until now, progeny testing has been a predominant method for the identification of the combining ability of genotypes that is most important for parental line selection. This approach is costly and time consuming.

Determination of genetic distances between parents was expected to predict the future hybrid performance. Several studies were conducted in maize, (*Zeya maya* L.), rape (*Brassica napus L.*), soybean (*Glycine maximum L.*), rice (*Oryza sativa L.*), wheat (*Triticum aestivum L.*) and other crops showing that the simple detection of relative genetic distance is not sufficient for such a purpose. Some genes were mapped controlling hybrid breakdown in rice and deleterious genes with synergistic interactions decreasing progeny fitness were described in wheat.

Their presence/absence in different genotypes might affect hybrid performance as well. Experiments were carried out in Arabidopsis to produce lines with better combining ability using recurrent selection. The research continues especially in corn.

Evaluation of the progeny follows. Whenever progeny is available, molecular markers can be used to help to select the best lines. The earliest investigations of Marker Assisted Selection (MAS) effectiveness on multiple loci selection were conducted in corn using multiple markers and the results were very promising. It has been shown until now that selection for traits encoded by a single gene is highly effective in several species using marker systems, especially when traditional evaluations are difficult, time consuming or expensive to run.

Marker-assisted selection can also accelerate the recovery of recurrent parent genome in backcross breeding and identify the plants with a higher relative proportion of recurrent parent genome for further backcrossing. Introgression of alien chromosome segments and effective chromosome markers can assist selection during the segregating backcross generations.

QTLs that determine important traits like quality, yield components and resistance to various stresses are expensive to develop and their practical application is rather limited at the moment. Usually it is not possible to use QTLs to evaluate other non-related crosses. Transfer of QTLs is also easier in backcrossing breeding programmes. The efficiency of marker-assisted selection for quantitative traits depends on the power of QTL detection and unbiased estimation of QTL effects. It has been shown that the accuracy of QTL location greatly affects selection efficiency.

Another important question of QTL mapping is the optimal choice of marker density. Whenever reliable markers are available, MAS can be used to pyramid several resistance genes into a single host genotype and to estimate the presence of quality and agronomically important gene blocks. At the moment, of course, it is necessary to combine conventional analyses of plant material with marker-assisted selection techniques. It is proposed that the use of marker data together with phenotypic evaluations provides instruments suitable for more effective breeding.

In case advanced lines are submitted for the state trials and approved for marketing, it is necessary to manage tools to identify the cultivars precisely—in seed lots, in products and processed food. At the moment, seed storage protein and isozyme analysis are recognised by international organisations as a regular tool for identification of plant varieties. Etalons were developed and published in special catalogues. However, there are several advantages of DNA fingerprinting over protein analysis. DNA analysis is currently more expensive than protein analysis, but it is possible to run it at any developmental stage of the plant and also in processed food and it covers the whole genome variability. It has been proved that molecular markers represent a fast and efficient tool to evaluate cultivar authenticity and purity.

Techniques for DNA Analysis

After the DNA structure and function were discovered, methods of DNA analysis expanded. From laborious and time consuming procedures used in the 70's and 80's laboratory protocols changed into pre-made semi-automated systems. The basic methodologies for marker development and application have also evolved.

Restriction Fragment Length Polymorphism (RFLP)

The method is based on the restriction endonuclease digestion of DNA and the transfer of DNA fragments to a filter where they can be hybridised by a labelled DNA fragment. Restriction endonucleases cut specific nucleotide motifs in a DNA sequence. The fragments have to be separated according to size in the gel by electrophoresis and the fragments of interest are identified by hybridisation to labelled probes.

A polymorphism in a restriction pattern occurs thanks to mutations such as single base-pair loss or gain mutations or mutations based on insertion/deletion, etc. RFLP patterns of nuclear DNA behave like classical codominant genetic markers and can be used to create RFLP linkage maps. The limitations of this method are that it is very labour intensive and expensive.

RFLP analysis is a well accepted method in plant breeding and is used for many different purposes: e.g. the selection of traits of agronomic importance linked to RFLP markers, quality testing of seeds and segregation analysis of progenies, evaluation of diversity in a germplasm collection. Molecular linkage maps based on RFLP markers were developed for major crop species including e.g. potato, maize and barley. RFLP was also used as a tool to describe the genetic variability of crop species.

Polymerase Chain Reaction (PCR)

The development of PCR technique is a milestone in genome analysis. The basic concept was tested for the first time with Klenow polymerase but the real breakthrough came when a thermostable DNA polymerase, Taq polymerase, was isolated and purified. PCR was originally conceived as a technique for detection base changes in the genome, as a tool for DNA diagnosis of genetic diseases. In the last few years, several assays to reveal DNA polymorphism at multialellic loci have been developed in the field of PCR. Considerable advantages of PCR-based methods are e.g. simplicity, speed, and specificity.

Random Amplified Polymorphic DNA (RAPD)

Arbitrarily Primed PCR (AP-PCR) and Randomly Amplified Polymorphic DNAs (RAPDs) are essentially the same technique. Most molecular biologists use more frequently the acronym RAPDs. RAPD technique requires only the presence of a single 'randomly chosen' oligonucleotide. Individual RAPD primers are able to hybridise to several hundred sites within the target DNA, however, not all of these hybridisations lead to the production of PCR fragments.

The ability of RAPDs to produce multiple bands using a single primer means that a relatively small number of primers can be used to generate a very large number of fragments. These fragments are usually generated from different regions of the genome and hence multiple loci may be examined very quickly. The power of RAPD is that it is a fast technique, easy to perform and comparatively cheap. It is immediately applicable to the analysis of most organisms because universal sets of primers are used without any need for prior sequence information.

This marker system was used in many different applications involving the detection of DNA sequence polymorphisms, mapping in different types of populations, isolation of markers linked to various traits or specific targeted intervals and applications such as variety identification and analysis of parentage.

The RAPD technology, however, has some limitations. RAPD markers are in general dominant, thereby they have a lower information content than codominant markers in the linkage analysis of F_2 populations. Another type of problem that has been reported is the occurrence of RAPD bands in progeny but not in their parental DNAs, a phenomenon explained as heteroduplex formation. It has been suggested that the outcome of RAPD reaction is in part determined by a competition for priming sites in the genome. In several mapping projects non-Mendelian inheritance for a significant fraction of all polymorphic bands was detected, possibly indicating problems with reproducibility and with competition.

Variable Number of Tandem Repeat Loci (VNTR)

The existence of microsatellite loci in eukaryotic genomes has been known since the 1970s. Tautz et al. showed that many of the simple sequences occurring in eukaryotes were 5 to 10 times more frequent than equivalent-sized random motifs, and that high numbers of 'cryptic' repeats or scrambled arrangements of repetitive sequences also occurred.

Jeffreys et al. discovered hypervariable tandem repeats in the human genome having a longer repeat unit (minisatellites). Minisatellites as well as microsatellites vary in the number of tandemly repeated elements, hence the general designation for both is a variable number of tandem repeat loci (VNTRs). VNTR analysis utilised the PCR, however only a limited subset of variations could be analysed by PCR due to the generally large sizes of minisatellite alleles. Microsatellites have the advantage of minisatellites because the allele sizes are smaller than 500 bp and the variation is over a narrow size range. Microsatellites have become the most important class of markers for linkage mapping in diverse organisms.

Simple Sequence Repeat (SSR)

Microsatellites consist of tandemly repeated units, each between one and 10 base-pairs in length, such as $(TG)_n$ or $(AAT)_n$. They are widely dispersed through eukaryotic genomes and are often highly polymorphic. These markers are one of the molecular tools of choice for biodiversity studies because of their high information content.

PCR amplification protocols used for microsatellites employ either unlabelled primer pairs or primer pairs with one of the primers being radiolabelled or fluorolabelled. Electrophoresis of unlabelled PCR products can be carried out on smaller vertical polyacrylamide gels or on horizontal agarose gels. However, this approach is not precise enough. Automated systems are also available. A major advantage of automated systems is the availability of dyes with different wavelengths, so it is possible to use simultaneous capillary electrophoresis of several loci with overlapping allele size ranges.

Scoring microsatellite gels or autoradiograms is usually a relatively simple process because the used electrophoresis systems have a high resolution (to a single base-pair) and because the alleles differ in a very predictable way (multiples of the microsatellite repeat unit, e.g. two base-pairs). The automated systems using fluorolabelled PCR products separated by capillary electrophoresis allowed to analyse these products using the software such as Genescan™ and Genotyper®. These analysis programmes provide algorithms that separate native alleles automatically from slippage products.

Microsatellites are co-dominant markers and the data generated are similar to those of allozymes, except that the number of alleles and heterozygosity revealed is almost always higher. Population genetic, parentage relatedness analysis can then be carried out. Slatkin and Goldstein et al. took advantage of knowledge of the predominant mode of microsatellite evolution (i.e. stepwise mutation) to derive the measures of population subdivision and average genetic distance (ASD).

Amplified Fragment Length Polymorphism (AFLP)

The principle of AFLP is based on a selectively amplifying a subset of restriction fragments from a complex mixture of DNA fragments obtained after digestion of genomic DNA with restriction endonucleases. Polymorphisms are detected from differences in the length of the amplified fragments by polyacrylamide gel electrophoresis (PAGE) or by capillary electrophoresis.

The technique involves four steps: (1) restriction of DNA and ligation of oligonucletide adapters; (2) preselective amplification; (3) selective amplification; (4) gel analysis of amplified fragments. Genomic DNA of an organism is digested with two different restriction enzymes, of which one has a 4-bp and the other a 6-bp recognition sequence. DNA isolated at first is digested with a pair of restriction enzymes usually recognising four and six nucleotide sequences, and adapters specific to the chosen restriction sites are ligated.

Amplification of restriction products follows. The selective amplification is achieved by the use of primers that extend into restriction fragments, amplifying only those fragments in which the primer extensions match the nucleotides flanking the restriction sites. One type of primer can be labelled, e.g. by fluorescent colour if the capillary electrophoresis system is used, or radioactive labelling can be used e.g. [γ-^{33}P] in the case of polyacrylamide electrophoresis. This method enables to visualise sets of restriction fragments by PCR without knowledge of nucleotide sequence.

The AFLP technology is a powerful tool for the detection and evaluation of genetic variation in germplasm collections and in the screening of biodiversity as well as for fingerprinting points. They can be successfully used for marker-assisted selection. Molecular markers linked to the Rfo restorer gene used for the Ogu-INRA cytoplasmic male-sterility system in rape, markers developed for linolenic acid content in rape, markers allowing selection for BaYMV resistance in barley, markers used to identify quantitative trait loci for grain yield and grain-related traits in maize, markers closely linked to the Rph7.g resistance gene of barley, DNA markers allowing marker-assisted breeding for Fusarium head blight resistance, DNA markers discriminating mutant and normal alleles at the Wx-D1 locus in wheat and some others. The application of the markers helps to speed up the breeding process and change some paradigms in plant breeding.

The above-mentioned techniques have been used to develop linkage maps of many plant species and DNA makers. However, the technical background makes it possible to employ still more effective approaches to genome characterisation.

Genomics

It has been proved that DNA markers could be useful for

characterisation of genetic resources, selection of parents and subsequently for easier screening of the best progeny or for identification of the genotypes to protect consumers or breeders in the market. However, only the precise knowledge of genome structure and function can lead to a better understanding of the genetic basis of superior genotypes and development of required culti-vars.

Therefore a new discipline called "genomics" started to develop. Genomics is a new field in biology that is concerned with the whole genome analysis, from sequence to function and derived information. The systematic analysis of plant genome function provides information on plant biology that will revolutionise plant and crop production. The crop species have been extensively studied, however, their genomes are quite large. More information is currently available about the model plant Arabidopsis thaliana, which is not of economic importance.

Arabidopsis thaliana, a crucifer, represents an important model system in plant molecular genetics. Due to its small genome, short generation time and high number of progeny, the plant is extremely suitable for genetic and mutation analyses. Likewise, the plant is ideally suited for molecular studies because with its content approximately 130 Mbp it is one of the smallest genomes known among higher plants.

The genome is characterised by a low content of repetitive sequences. Large collections of partial cDNA sequences are also available and the entire sequence of the nuclear genome was deciphered. All these materials and information are accessible through databases as well as DNA and seed stock centres. Once the genome structure is recognised, the genome function is investigated. There are several approaches of the function of plant genome—(1) Gene expression analysis following an exposure of plants to stresses and comparative analysis is employed to identify functional isoforms of genes, (2) mutants —stock of hundreds of mutant lines is available and new ones are being developed using T-DNA tagging. A new system for insertional mutagenesis based on the maize Enhancer/Suppressor-mutator (En/Spm) element was also introduced into *Arabidopsis thaliana*. Recent progress in large-scale insertional mutagenesis opens new possibilities of functional genomics in Arabidopsis.

The number of T-DNA and transposon insertion lines from different laboratories will soon represent insertions into most *Arabidopsis* genes. Vast resources of gene knockouts are becoming available that can be subjected to different types of reverse genetics screens to deduce the functions of sequenced genes. With rapid progress in the genome projects of different plants, large-scale transposon mutagenesis has become an important component of functional genomics, permitting assignment of functions to sequenced genes through reverse genetics.

Knockouts of genes encoding enzymes of primary metabolism can produce mutants with clear and sometimes unexpected phenotypes. Genome data have to be converted into knowledge to be useful to biologists. Many valuable computational tools have already been developed to help annotation of plant genome sequences, and they may be improved in future.

Comparative genomics is another possibility of gene discoveries. The genomic structure of *Arabidopsis thaliana* is compared with animal and microbial genomes to sense the function of some of the genomic regions. In fact the information on *Arabidopsis* structure can lead to discovery of genes of other plant species. The first homoeologous segments identified in the genomes of a dicot and monocot demonstrate that the fine-scale conservation of genome structure exists and is detectable across the angiosperms. Comparative sequencing studies reveal higher degrees of diversity at the microstructural (less than 1 million base pairs) level than predicted at the genetic map level and suggest that genes are densely packed in gene-rich regions. The conserved framework of identified genes is interspersed with non-conserved genes, which however indicates that the mechanisms beyond segmental inversions and translocations need to be invoked to fully explain the plant genome evolution, and that the benefits of comparative genomics over such large taxonomic distances may be limited. Besides comparative DNA studies metabolite profiling can also be a new tool for a comparative display of gene function. It has the potential not only to provide deeper insight into complex regulatory processes but also to determine the phenotype directly.

Model plant *Arabidopsis* provide knowledge of the function of plant genes with unprecedented clarity and quantity. Comparative genetic mapping experiments established colinearity of genomes for the species of the Brassicaceae. While *Arabidopsis thaliana* is a model for dicotyledons, rice has been selected as a model plant for monocotyledonous families because of its relatively small genome, conservative genome organisation among the cereals and global use of rice.

Sequence comparisons between *Arabidopsis* and rice can also define some potential functional relationships, and the information can be used to ascribe functions to genes in many cereals. A major challenge now is to apply this new information to the improvement of crop plants in a systematic manner. Similar techniques are used for rice study including T-DNA mutagenesis. Grasses are the most important plant family in agriculture.

Comparative genetic mapping has revealed the conserved gene order (colinearity) between many grass species. It was demonstrated however that the micro-colinearity of genes is less conserved: small-scale rearrangements and deletions complicate the microcolinearity between closely related species. Therefore rice have to be complemented by more intensive genetic work on grass species with large genomes (maize, Triticale). The large genome of barley and wheat functional genomic approaches, focused on the expressed portion of the genome, have recently led to an exponential growth of expressed sequence tagged (EST) databases of cereals. Assigning gene function to these ESTs is now one of the major challenges in wheat genomics.

5

CHARACTERISTICS OF TRANSGENIC PLANTS

Transgenic plants are plants that have been genetically engineered, a breeding approach that uses recombinant DNA techniques to create plants with new characteristics. Transgenic plants are produced by adding one or more genes to a plant's genome, by a process called transformation. They are identified as a class of genetically modified organism (GMO). Transgenic plants have been developed for a variety of reasons: longer shelf life, disease resistance, herbicide resistance, and pest resistance. The first transgenic crop approved for sale in the US, in 1994, was the FlavrSavr tomato, which was intended to have a longer shelf life.

Today, there are more than 10,677,000 km² of transgenic plants being grown throughout the world. There are three general types of transgenic plants: those with genes to improve the quality of the product, those with genes to allow them to resist disease or herbivory (consumption by herbivores, usually insects), and plants with genes that allow them to be resistant to the effects of specific herbicides. Transgenic crops are grown world wide, although the greatest concentration of transgenic crops is in the United States.

An emerging class of transgenic plant, sometimes known as pharmacrops, aims to use plants to manufacture other products, such as pharmaceuticals and industrial chemicals. Testing of a variety of these crops has been underway for several years. It appears that no such products have yet been used in the commercial market. Cultivation of transgenic crops is a highly controversial issue, and a primary target of anti-GMO activists. Food safety and potential ecological hazards are central concerns.

Traditional Breeding and Transgenic Technology

For several thousand years, farmers have been altering the genetic makeup of the crops they grow. Human selection for features such as faster growth, larger seeds or sweeter fruits has dramatically changed domesticated plant species compared to their wild relatives. Remarkably, many of our modern crops were developed by people who lacked an understanding of the scientific basis of plant breeding. Despite the poor understanding of the process, plant breeding was a popular activity.

Gregor Mendel himself, the father of genetics, was a plant breeder, as were some of the leading botanists of his time. Mendel's 1865 paper explaining how dominant and recessive alleles could produce the traits we see and could be passed to offspring was the first major insight into the science behind the art. The paper was largely ignored until 1900, when three scientists working on breeding problems rediscovered it and publicised Mendel's findings. Major advances in plant breeding followed the revelation of Mendel's discovery. Breeders brought their new understanding of genetics to the traditional techniques of self-pollinating and cross-pollinating plants.

Corn breeders, particularly, tried numerous strategies to capitalise on the insights into heredity. Corn plants that had traditionally been allowed to cross-pollinate freely were artificially self-pollinated for generations and crossed to other self-pollinated lines in an effort to achieve a favourable combination of alleles. The corn we eat today is the result of decades of this strategy of self-pollination followed by cross-pollination to produce vigorous hybrid plants. The art of recognising valuable traits and incorporating them into future generations is very important in plant breeding.

Breeders have traditionally scrutinised their fields and travelled to foreign countries searching for individual plants that exhibit desirable traits. Such traits occasionally arise spontaneously through a process called mutation, but the natural rate of mutation is too slow and unreliable to produce all the plants that breeders would like to see. In the late 1920s, researchers discovered that they could greatly increase the number of these variations or mutations by exposing plants to X-rays. "Mutation breeding" accelerated after World War II, when the techniques of the nuclear age became widely available.

Plants were exposed to gamma rays, protons, neutrons, alpha particles, and beta particles to see if these would induce useful mutations. Chemicals, too, such as sodium azide and ethyl methanesulphonate, were used to cause mutations. Mutation breeding was particularly popular in the United States during the 1970s. Although interest has waned somewhat in recent years, occasional varieties continue to be produced using these methods. Mutation breeding efforts continue around the world today. Of the 2,252 officially released mutation breeding varieties, 1,019, or almost half, have been released during the last 15 years.

Another method for increasing the number of mutations in plants is tissue culture. Tissue culture is a technique for growing cells, tissues, and whole plants on artificial nutrients under sterile conditions, often in small glass or plastic containers. Tissue culture was not developed with the intention of causing mutations, but the discovery that plant cells and tissues grown in tissue culture would mutate rapidly increased the range of methods available for mutation breeding. It was during the 1970s also that haploid breeding was heavily utilised.

Spontaneously occurring haploid plants, those having half the normal number of chromosomes, were discovered in the

1920s, but haploid breeding was not a practical technique until methods for the controlled production of haploid plants were developed. Once a haploid plant has been obtained, its chromosomes are artificially doubled to return the plant to the normal number of chromosomes. Such a plant is valuable because the chromosomes that were created by artificial doubling are exact copies of the chromosomes that were present in the haploid plant. Haploids have been used in creating cultivars of barley, maize, tobacco, asparagus, strawberries, and tall fescue grass. Often these plants are more useful in basic research than in commercial applications, but the haploid-derived barley cultivar Tangangara was released for commercial production in Australia in 1996.

While most breeders cross-pollinate plants of a single species, some breeding methods rely on crosses that can be made between two species within the same genus. A cross between Musa acuminata and Musa balbisiana, both members of the genus Musa, produced the bananas with which we are familiar. Less commonly, the cross is between members of two different genera. A cross between wheat, Triticum aestivum, and rye, Secale cereale, produced the grain called triticale, which contains a copy of all the chromosomes from both species. A variation on the wide crossing procedure is to select plants that have single chromosomes or chromosome arms substituted from one species into another.

Many modern wheat cultivars, for example, contain a chromosome arm from rye, which adds resistance to several diseases. Transgenic technology provides the means to make even more distant "crosses" than were previously possible. Organisms that have until now been completely outside the realm of possibility as gene donors can be used to donate desirable traits to crop plants. These organisms do not provide their complete set of genes, but rather donate only one or a few genes to the recipient plant. For example, a single insect-resistance gene from the bacterium Bacillus thuringiensis can be transferred to a corn plant to make Bt corn.

Benefits of GM Crops

Economical

GM supporters tell farmers that they stand to reap enormous profits from growing GM crops. Initially, the cost is expensive but money is saved on pesticides. To produce the GM crops, modern biotechnology is used which requires highly skilled people and sophisticated and expensive equipment. Large companies need considerable investments in laboratories, equipment and human resources, hence the reason why GM crops are more expensive for farmers than traditional crops. GM crops, farmers are told, are a far better option. It takes a shorter time to produce the desired product, it is precise and there are no unwanted genes.

Herbicide-resistant crops

GM crops can be produced to be herbicide resistant. This means that farmers could spray these crops with herbicide and kill the weeds, without affecting the crop. In effect, the amount of herbicide used in one season would be reduced, with a subsequent reduction in costs for farmers and consumers. For Ingard cotton, pest resistance was built into the cotton, hence reducing and even removing the use of pesticides, which are not only expensive but, more importantly, harmful to the environment. Biotechnology companies are even experimenting with crops that can be genetically modified to be drought and salt-tolerant, or less reliant on fertilizer, opening up new areas to be farmed and leading to increased productivity. However, the claims of less herbicide usage with GM crops have till now not been independently supported by facts.

Better quality foods

Even animals can be genetically modified to be leaner, grow faster, and need less food. They could be modified to have special characteristics, such as greater milk production in cows.

These modifications again lead to improved productivity for farmers and ultimately lower costs for the consumer. Modified crops could perhaps prevent outbreaks such as foot and mouth disease, which has devastated many farmers and local economies.

No such products have been released to date; however, some are under consideration for release. For example, GM salmon, capable of growing almost 30 times faster than natural salmon, may soon be approved by the FDA (Food and Drug Administration) in the U.S.. The following are some examples of food plants that are undergoing field trials:

— apples that resist insect attack

— bananas free of viruses and worm parasites

— coffee with a lower caffeine content

— cabbage that resists caterpillar attacks

— melons that have a longer shelf life

— sunflowers that produce oil with lower saturated fat

Methods for Constructing Transgenic Plant

Introduction to DNA

The underlying reason that transgenic plants can be constructed is the universal presence of DNA (deoxyribonucleic acid) in the cells of all living organisms. This molecule stores the organism's genetic information and orchestrates the metabolic processes of life. Genetic information is specified by the sequence of four chemical bases (adenine, cytosine, guanine, and thymine) along the length of the DNA molecule. Genes are discrete segments of DNA that encode the information necessary for assembly of a specific protein.

The proteins then function as enzymes to catalyse biochemical reactions, or as structural or storage units of a cell, to contribute to expression of a plant trait. The general sequence of events by which the information encoded in DNA is expressed in the form of proteins via an mRNA intermediary is shown below.

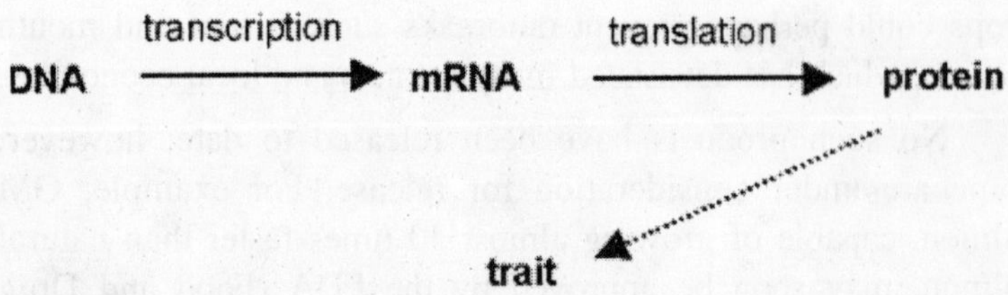

The transcription and translation processes are controlled by a complex set of regulatory mechanisms, so that a particular protein is produced only when and where it is needed. Even species that are very different have similar mechanisms for converting the information in DNA into proteins; thus, a DNA segment from bacteria can be interpreted and translated into a functional protein when inserted into a plant. Among the most important tools in the genetic engineer's tool kit are enzymes that perform specific functions on DNA.

Identifying and Locating Genes

Identifying and locating genes for agriculturally important traits is currently the most limiting step in the transgenic process. We still know relatively little about the specific genes required to enhance yield potential, improve stress tolerance, modify chemical properties of the harvested product, or otherwise affect plant characters. Usually, identifying a single gene involved with a trait is not sufficient; scientists must understand how the gene is regulated, what other effects it might have on the plant, and how it interacts with other genes active in the same biochemical pathway.

Designing Genes

Once a gene has been isolated and cloned (amplified in a bacterial vector), it must undergo several modifications before it can be effectively inserted into a plant. Simplified representation of a constructed transgene, containing necessary components for successful integration and expression.

1. A promoter sequence must be added for the gene to be correctly expressed (i.e., translated into a protein product). The promoter is the on/off switch that controls when and where in the plant the gene will be expressed. To date, most promoters in transgenic crop varieties have been "constitutive", i.e., causing gene expression throughout the life cycle of the plant in most tissues. The most commonly used constitutive promoter is CaMV35S, from the cauliflower mosaic virus, which generally results in a high degree of expression in plants. Other promoters are more specific and respond to cues in the plant's internal or external environment. An example of a light-inducible promoter is the promoter from the cab gene, encoding the major chlorophyll a/b binding protein.

2. Sometimes, the cloned gene is modified to achieve greater expression in a plant. For example, the Bt gene for insect resistance is of bacterial origin and has a higher percentage of A-T nucleotide pairs compared to plants, which prefer G-C nucleotide pairs. In a clever modification, researchers substituted A-T nucleotides with G-C nucleotides in the Bt gene without significantly changing the amino acid sequence. The result was enhanced production of the gene product in plant cells.

3. The termination sequence signals to the cellular machinery that the end of the gene sequence has been reached.

4. A selectable marker gene is added to the gene "construct" in order to identify plant cells or tissues that have successfully integrated the transgene. This is necessary because achieving incorporation and expression of transgenes in plant cells is a rare event, occurring in just a few percent of the targeted tissues or cells. Selectable

marker genes encode proteins that provide resistance to agents that are normally toxic to plants, such as antibiotics or herbicides. Only plant cells that have integrated the selectable marker gene will survive when grown on a medium containing the appropriate antibiotic or herbicide. As for other inserted genes, marker genes also require promoter and termination sequences for proper function.

Plant Transformation

Transformation is the heritable change in a cell or organism brought about by the uptake and establishment of introduced DNA. There are two main methods of transforming plant cells and tissues:

1. The "Gene Gun" method (also known as microprojectile bombardment or biolistics).

2. The Agrobacterium method, which is described below. Transformation via Agrobacterium has been successfully practiced in dicots (broadleaf plants like soybeans and tomatoes) for many years, but only recently has it been effective in monocots (grasses and their relatives). In general, the Agrobacterium method is considered preferable to the gene gun, because of the greater frequency of single-site insertions of the foreign DNA, making it easier to monitor.

Agrobacterium method

Agrobacterium tumefaciens is a remarkable species of soil-dwelling bacteria that has the ability to infect plant cells with a piece of its DNA. When the bacterial DNA is integrated into a plant chromosome, it effectively hijacks the plant's cellular machinery and uses it to ensure the proliferation of the bacterial population. Many gardeners and orchard owners are unfortunately familiar with A. tumefaciens, because it causes crown gall diseases in many ornamental and fruit plants.

The DNA in an A. tumefaciens cell is contained in the bacterial chromosome as well as in another structure known as a Ti (tumor-inducing) plasmid. The Ti plasmid contains

— a stretch of DNA termed T-DNA (~20 kb long) that is transferred to the plant cell in the infection process.

— a series of vir (virulence) genes that direct the infection process.

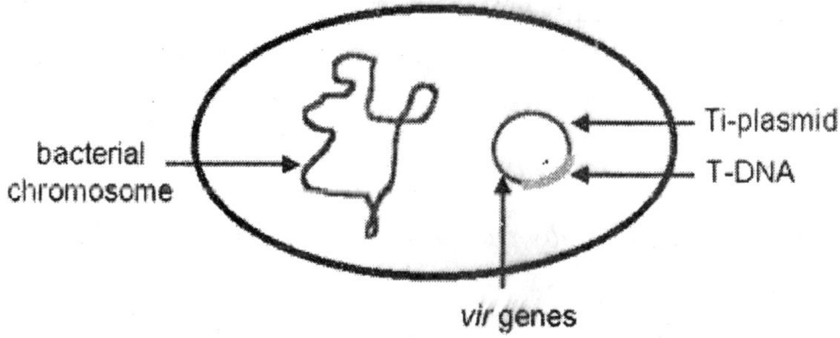

Agrobacterium tumefaciens cell

A tumefaciens can only infect a plant through wounds. When a plant root or stem is wounded it gives off certain chemical signals. In response to those signals, the vir genes of A. tumefaciens become activated and direct a series of events necessary for the transfer of the T-DNA from the Ti plasmid to the plant's chromosome. Different vir genes

— Copy the T-DNA.

— Attach a product to the copied T-DNA strand to act as a leader.

— Add proteins along the length of the T-DNA, possibly as a protective mechanism.

— Open a channel in the bacterial cell membrane, through which the T-DNA passes.

The T-DNA then enters the plant cell through the wound. It is not clear how the bacterial DNA moves from the cytoplasm to the nucleus of the plant cell, nor how the T-DNA becomes integrated into the plant chromosome. Remember that most of the time plant DNA does not exist as an exposed strand, but is wrapped with histone proteins and is in a supercoiled state. One speculation is that the T-DNA waits until the plant DNA is being replicated or transcribed, then inserts itself into the exposed plant DNA. To harness A. tumefaciens as a transgene vector, scientists have removed the tumor-inducing section of T-DNA, while retaining the T-DNA border regions and the vir genes. The transgene is inserted between the T-DNA border regions, where it is transferred to the plant cell and becomes integrated into the plant's chromosomes.

Selection and Regeneration

Selection of successfully transformed tissues: Following the gene insertion process, plant tissues are transferred to a selective medium containing an antibiotic or herbicide, depending on which selectable marker was used. Only plants expressing the selectable marker gene will survive and it is assumed that these plants will also possess the transgene of interest. Thus, subsequent steps in the process will only use these surviving plants.

Regeneration of whole plants: To obtain whole plants from transgenic tissues such as immature embryos, they are grown under controlled environmental conditions in a series of media containing nutrients and hormones, a process known as tissue culture. Once whole plants are generated and produce seed, evaluation of the progeny begins. This regeneration step has been a stumbling block in producing transgenic plants in many species, but specific varieties of most crops can now be transformed and regenerated.

Plant Breeding and Testing

Intrinsic to the production of transgenic plants is an extensive evaluation process to verify whether the inserted gene has been stably incorporated without detrimental effects to other plant functions, product quality, or the intended agroecosystem. Initial evaluation includes attention to:.

— Activity of the introduced gene

— Stable inheritance of the gene

— Unintended effects on plant growth, yield, and quality

If a plant passes these tests, most likely it will not be used directly for crop production, but will be crossed with improved varieties of the crop. This is because only a few varieties of a given crop can be efficiently transformed, and these generally do not possess all the producer and consumer qualities required of modern cultivars. The initial cross to the improved variety must be followed by several cycles of repeated crosses to the improved parent, a process known as backcrossing. The goal is to recover as much of the improved parent's genome as possible, with the addition of the transgene from the transformed parent.

Environmental Impacts of Transgenic Plants

Transgenes have the potential for significant ecological impact if the plants can increase in frequency and persist in natural populations. This can occur:

— if the transgene escapes from the fields into the wild.

— if transgenic plants mate with similar wild plants, the transgene could be incorporated into the offspring.

— if these new transgene plants become weedy or invasive, which could reduce biodiversity and might destroy entire ecosystems.

Transgene Escape

Escape can happen several different ways. First, the transgenic seeds can end up being carried out of the field by wind or animals, or can be spread through volunteer plants, which grow up from the seeds dropped by the previous year's harvest. Second, transgenic plants in the field can hybridize with neighboring wild relatives, and spread the transgene that way. This is the method of transgenic plant escape that has the greatest potential to spread the transgene.

The reason that hybridization is so much more worrisome is that it has the potential to create so many more transgenic plants. This is because of the large number of seeds produced by most crop plants. There are three possible avenues of hybridization:

1. Hybridization with non-transgenic crop plants of the same species and variety.
2. Hybridization with wild plants of the same species.
3. Hybridization with wild plants of closely related species, usually of the same genus.

However, there are a number of factors which must be present for hybrids to be created.

— The transgenic plants must be close enough to the wild species for the pollen to reach the wild plants.

— The wild and transgenic plants must flower at the same time.

— The wild and transgenic plants must be genetically compatible.

— The hybrid offspring must be viable, and fertile.

— The hybrid offspring must carry the transgene.

Transgenic and non-transgenic plants of the same species only have to overcome the first barrier in order to produce hybrids because they are the same species and therefore interbreed easily. Some transgenic species are aiso fertile with their closest wild relative, which is generally the same species as the cultivated and transgenic plants. Thus where ever transgenic plants are grown along side wild populations of the same species, hybridization is virtually guaranteed. This interbreeding between wild and cultivated plants has been seen in sunflowers, oil-seed rape (Brassica napus), rice and all varieties of beets, as well as a number of other species.

Hybridization with closely related species is less common. Some crops, such as rice, can hybridize with other species in the same genus. However, cultivated sunflowers (Helianthus annuus) do not interbreed with other species of sunflower (H. petiolris) very well, and oil-seed rape (Brassica napus) does not hybridize with the closely related weed wild mustard (Sinapis arvenis) except under laboratory conditions.

Transgene Expression

Studies suggest that the most likely escape route for transgenic plants will be through hybridization with wild plants of the same species. To address this concern, we need to evaluate the phenotype, or physical characteristics, of the hybrid plants. If the transgenic trait is expressed in the hybrid plants, we will be able to see evidence of it over the life cycle of the plant. For a hybrid plant to express the transgene it must first receive the gene from one of its parents.

Using genetic testing such as PCR scientists can determine if a plant has the gene in question by looking for genetic markers. The next question to ask is whether the hybrid plant expresses the gene. This is tested by looking at what the gene is supposed to do in the plant. For example, in the case of the oilseed rape plants with multiple herbicide resistances, seedlings were tested for herbicide resistance by applying a small amount of the herbicide in question to one leaf. Resistant plants were not affected, while non-resistant plants had wilted leaves. While the exact rate of hybridization is not known, approximately three out of four hybrid plants carried the transgene.

In another experiment sunflowers with transgenic resistance to wild mold were hybridized with wild sunflowers. The resulting offspring were examined genetically for the resistance gene and then exposed to white mold to judge their resistance. In this case half of the hybrid plants carried the transgene. There have been similar experiments with Bt pesticides in oilseed rape and Bt sunflowers. These sunflowers were also found to produce hybrids where 49.3% carried the Bt gene.

In addition to enetic testing for the transgene, the sunflowers were also shown to express the gene by having significantly less insect damage than the wild sunflowers. This means that the transgene continues to be effective in hybrid plants. The evidence shows that transgenic hybrids can be expected to have and express the transgene. So we know that transgenic hybrids will exist, and at least some will express the transgene.

Transgene Fitness

It is very important to consider if the transgenic plants have an advantage over other plants in the wild. If transgenic plants do not have any survival advantage due to the transgene, they are not at risk of becoming weedy or invasive. While this question is vital, it has not been studied extensively. There are, however, a set of generally accepted hypotheses about the type of trait that will and will not lead to increased weedyness.

1. Traits which confer a significant advantage over other plants with respect to non-living (abiotic) factors, such as drought and salt tolerance will increase rapidly in the population, and could cause the plant to become weedy or invasive. At this point there are crop plants with some tolerance to salt, however none of these plants are transgenic.

2. Traits which are only advantageous under selection pressure are likely to increase where ever there is selection for that trait, usually insect, disease, or herbicide resistance. The frequency of these traits when there is no advantage to them will depend on the cost of the trait, and genetic drift.

   — The trait for Bt-enabled insect resistance is one such trait that should confer higher fitness when the plants are regularly stressed by insect attack. In an experiment involving the survivorship and reproductive output of two lines of transgenic oilseed rape (rapeseed, canola, Brassica napus) with Bt resistance to insects, plants with the transgene suffered less damage from the insects, and plants with less insect damage were more likely to survive the winter and produced more seeds than the non-transgenic plants, which had serious insect damage. Thus the transgenic plants had greater fitness than the non-transgenic plants.

— However, in a similar experiment involving hybrid sunflowers with a transgene (OxOx) which confers a resistance to white mold, the transgenic plants were found to have very little advantage over the wild plants. The transgene did not always confer resistance to the white mold, and resistance did not significantly alter the amount of seeds produced. Because the reproductive output was the same, the transgenic plants had the same fitness as the wild plants. It has been suggested that this resistance was not helpful to the hybrid plants because the wild sunflowers already have some kind of resistance.

3. Traits that reduce the plant's fitness (ability to survive and produce fertile offspring) will not become more common in the population and will not become weedy or invasive. This would include plants with genes for longer shelf life, greater nutritional content, or pharmaceuticals, all of which are costly for the plant to produce, but do not improve its survival.

Preventing Transgene Escape

At this time the containment of transgenic crops is generally limited to the same physical separation used for most crops. However other methods are being developed to prevent the spread of the transgene itself. For the most part these consist of developing transgenic plants which are sterile, and so cannot spread the transgene through reproduction. However, this method is not foolproof, and is totally useless when the desired product is the reproductive part of the plant, which is the case for almost all transgenic plants. It also creates the issue of unfair supply control by the companies which produce the seed, since new seed would have to be bought every year, a serious problem for poor farmers.

Other suggested methods would limit the transgene to the particular tissues to be altered. Most of the proposed methods are still in the early stages of research and development. At the present time the best method for containing transgenic plants looks to be simple physical containment, by planting the transgenic crop far from any other plants it might hybridize with.

Bt-Resistant Crops

"Bt" is short for Bacillus thuringiensis, a soil bacterium whose spores contain a crystalline (Cry) protein. In the insect gut, the protein breaks down to release a toxin, known as a delta-endotoxin. This toxin binds to and creates pores in the intestinal lining, resulting in ion imbalance, paralysis of the digestive system, and after a few days, insect death. Different versions of the Cry genes, also known as "Bt genes", have been identified. They are effective against different orders of insects, or affect the insect gut in slightly different ways. The use of Bt to control insect pests is not new.

Insecticides containing Bt and its toxins (e.g., Dipel, Thuricide, Vectobac) have been sold for many years. Bt-based insecticides are considered safe for mammals and birds, and safer for non-target insects than conventional products. What is new in Bt crops is that a modified version of the bacterial Cry gene has been incorporated into the plant's own DNA, so that the plant's cellular machinery produces the toxin. When the insect chomps on a leaf or bores into a stem of a Bt-containing plant, it ingests the toxin and will die within a few days. The use of

Bt varieties has dramatically reduced the amount of chemical pesticides applied to cotton.

Although Bt genes have proven to be quite effective in the short term for protecting against crop insect damage, as well as reducing fungal contamination of corn, there are concerns that widespread use of Bt varieties will accelerate development of resistance to Bt in the target pests. This could mean the loss of Bt as an effective, environmentally friendly insecticide.

Beginning in 2000, the EPA requires that farmers growing Bt corn must plant at least 20% of their total corn acreage to a non-Bt variety. The rationale is that the few Bt-resistant insects surviving in the Bt field would likely mate with susceptible individuals that have matured in the non-Bt refuge. Thus, the insect genes (alleles) for resistance to Bt would be swamped by the susceptible alleles. Whether this strategy will work or not remains to be seen. Some of the potential problems with the refuge strategy are:

— The frequency of Bt-resistant alleles in insect populations may be greater than assumed in refuge models.

— Resistance to Bt in European corn borer may be semi-dominant rather than recessive.

— Resistant insects surviving in the Bt field may mature several days later than susceptible insects in the refuge, thus preventing their mating.

Negative Consequences of GM Crops

The introduction of transgenic crops and foods into the existing food production system has generated a number of questions about possible negative consequences. People with concerns about this technology have reacted in many ways, from participating in letter-writing campaigns to demonstrating in the streets to vandalizing institutions where transgenic research is being conducted. The issues surrounding objections to transgenic crops can be broadly grouped into concerns about

1. Damage to human health
   — allergenicity
   — horizontal transfer and antibiotic resistance
   — eating foreign DNA
   — cauliflower mosaic virus promoter
   — changed nutrient levels
2. Damage to the natural environment
   — Monarch butterfly
   — crop-to-weed gene flow
   — antibiotic resistance
   — leakage of GM proteins into soil
   — reductions in pesticide spraying: are they real?
3. Disruption of current practices of farming and food production in developed countries
   — crop-to-crop gene flow
4. Disruption of traditional practices and economies in less developed countries.

Damage to Human Health

Allergenicity

The possibility that we might see an increase in the number of allergic reactions to food as a result of genetic engineering has a powerful emotional appeal because many of us experienced this problem before the advent of transgenic crops, or know of someone who did. However, there is no evidence so far that genetically engineered foods are more likely to cause allergic reactions than are conventional foods. Tests of several dozen transgenic foods for allergenicity have uncovered only a soybean that was never marketed and the now-famous StarLink corn. Although the preliminary finding is that StarLink corn is probably not allergenic, the scientific debate continues.

Every year some people discover that they have developed an allergy to a common food such as wheat or eggs, and some people may develop allergies to transgenic foods in the future, but there is no evidence that transgenic foods pose more of a risk than conventional foods do.

Horizontal transfer and antibiotic resistance

The use of antibiotic resistance markers in the development of transgenic crops has raised concerns about whether transgenic foods will play a part in our loss of ability to treat illnesses with antibiotic drugs. At several stages of the laboratory process, developers of transgenic crops use DNA that codes for resistance to certain antibiotics, and this DNA becomes a permanent feature of the final product although it serves no purpose beyond the laboratory stage.

Antibiotic pills

Transfer of a resistance gene from transgenic food to micro-organisms that normally inhabit our stomach and intestines, or to bacteria that we ingest along with food, could help those micro-organisms to survive an oral dose of antibiotic medicine. Although horizontal transfer of DNA does occur under natural circumstances and under laboratory conditions, it is probably quite rare in the acid environment of the human stomach. Another concern is that the enzyme product of the DNA might be produced at low levels in transgenic plant cells. While high processing temperatures would inactivate the enzyme in processed foods, ingestion of fresh or raw transgenic foods could result in the stomach containing a small amount of an enzyme that inactivates an orally administered dose of the antibiotic.

Eating foreign DNA

When scientists make a transgenic plant, they insert pieces of DNA that did not originally occur in that plant. Often these pieces of DNA come from entirely different species, such as viruses and bacteria. We eat DNA every time we eat a meal. DNA is the blueprint for life and all living things contain DNA in many of their cells. What happens to this DNA? Most of it is broken down into more basic molecules when we digest a meal.

A small amount is not broken down and is either absorbed into the blood stream or excreted in the feces. We suspect that the body's normal defense system eventually destroys this DNA. Further research in this area would help to determine exactly how humans have managed to eat DNA for thousands of years without noticing any effects from the tiny bits that sneak into the bloodstream. So far there is no evidence that DNA from transgenic crops is more dangerous to us than DNA from the conventional crops, animals, and their attendant micro-organisms that we have been eating all our lives.

Cauliflower mosaic virus promoter (CaMV promoter)

When scientists use transgenic technology to put a new gene into a plant, they put in additional pieces of DNA to direct the activity of that gene. One of these pieces is the "promoter" that turns the gene on. The most widely used promoter is the cauliflower mosaic virus 35S promoter, often abbreviated as the CaMV promoter or the 35S promoter. This promoter was obtained from the virus that causes cauliflower mosaic disease in several vegetables, such as cauliflower, broccoli, cabbage, and canola.

There are concerns that the CaMV promoter might be harmful if it were to invade our cells and turn on our genes. A multi-step chain of events would have to occur for the CaMV promoter to escape the normal digestive breakdown process, penetrate a cell of the body, and insert itself into a human chromosome. While there have been no tests to determine whether the CaMV promoter has invaded human tissues, experiments with mice indicate that normal body defenses eliminate stray fragments of foreign DNA that sneak into the blood stream from the digestive tract.

There is some evidence that the CaMV promoter poses little threat to human health. People have been eating it in small quantities for hundreds of years when we eat vegetables that are infected with the disease. Although vegetables heavily infected with CaMV are unappetizing, there have been no documented negative effects on health from eating the virus or its promoter.

Changed nutrient levels

How do genetically engineered foods compare with conventional foods in nutritional quality? This is an important issue, and one for which there will probably be much research in the future, as crops that are engineered specifically for improved nutritional quality are marketed. The central question for GE crops that are currently available is whether plant breeders have accidentally changed the nutritional components that we associate with conventional cultivars of a crop. Because isoflavones are thought to play a role in preventing heart disease, breast cancer, and osteoporosis, the isoflavone content of RoundupReady soybeans has been investigated by several researchers.

Damage to the Environment

Monarch butterfly

The suggestion that Bt corn pollen might kill Monarch butterfly larvae galvanised public interest in the effect of transgenic crops on the environment..

Crop-to-weed gene flow

Hybridization of crops with nearby weeds may enable weeds to acquire traits we wish they didn't have, such as resistance to herbicides. Research results indicate that crop traits may escape from cultivation and persist for many years in wild populations. Genes that provide a competitive edge, such as resistance to viral disease, could benefit weed populations around a crop field.

Many cultivated crops have sexually compatible wild relatives with which they hybridize under favourable circumstances.

The likelihood that transgenes will spread can be different for each crop in each area of the world. Soybeans and wheat are self-pollinating crops, so the risk of transgenic pollen moving to nearby weeds is small. However, that small risk must balanced against the fact that there are wild relatives of wheat in the United States. There are no wild relatives of soybean in the United States, but such wild relatives exist in China. Thus each crop must be evaluated individually for the risk of gene flow in the area where it will be grown.

Antibiotic resistance

There is also concern that transgenic plants growing in the field will transfer their antibiotic resistance genes to soil micro-organisms, thus causing a general increase in the level of antibiotic resistance in the environment. However, many soil organisms have naturally occurring resistance as a defense against other organisms that generate antibiotics, so genes contributed occasionally by transgenic plants are unlikely to cause a change in the existing level of antibiotic resistance in the environment.

Leakage of GM proteins

Many plants leak chemical compounds into the soil through their roots. There are concerns that transgenic plants may leak different compounds than conventional plants do, as an unintended consequence of their changed DNA. Speculation that this may be happening leads to concern about whether the communities of micro-organisms living near transgenic plants may be affected.

The interaction between plants and soil micro-organisms is very complex, with the micro-organisms that live around plant roots also leaking chemical compounds into the soil. Much more research must be done before we understand the relationships that occur between micro-organisms and conventional crops.

Reductions in pesticide spraying

One of the most appealing arguments in favour of transgenic plants is the potential for reducing the damage we do to our environment with conventional methods of farming. Pest-resistant crops such as Bt corn and Bt cotton have been promoted as a means to reduce the spraying of pesticides, while herbicide-tolerant crops such as RoundupReady soybeans are said to reduce the application of herbicides. Large reductions in chemical spraying have been claimed to result from the introduction of these transgenic varieties. Bt cotton is the only crop for which claims of reduced spraying are clear. Analysts paint a mixed picture on the results of planting RoundupReady soybeans. Bt corn and herbicide-tolerant cotton and corn have not resulted in clear reductions in the spraying of chemicals.

Damage to Current Farming Practices

Crop-to-crop gene flow

Hybridization of transgenic crops with nearby conventional crops raises concerns about separation distances to ensure purity of crops and about who must pay if unwanted genes move into a neighbour's crop. As "Identity Preservation" and segregation of GM from non-GM crops become factors in marketing products, it will be important to ensure that hybridization is not occurring in the field. Many factors influence the potential for gene flow from crop to crop.

Some crops are highly outcrossing, with pollen carried to other fields by wind and by insects. Other species are highly self-pollinating, with little potential for pollen transfer to neighboring plants. Because of the differences among crops species, every case must be evaluated individually for potential to contribute to gene flow from transgenic to conventional crops. If GM pollen pollinates plants in a neighboring field, then the issue of genetic trespass may arise.

FUTURE DEVELOPMENTS

New techniques for producing transgenic plants will improve the efficiency of the process and will help resolve some of the environmental and health concerns. Among the expected changes are the following:

— More efficient transformation, that is, a higher percentage of plant cells will successfully incorporate the transgene.

— Better marker genes to replace the use of antibiotic resistance genes.

— Better control of gene expression through more specific promoters, so that the inserted gene will be active only when and where needed.

— Transfer of multi-gene DNA fragments to modify more complex traits.

Future Transgenic Products

Tomato

Because tomatoes are one of the world's most popular vegetables, they have benefited from a long history of genetic improvement that continues in the transgenic age. Lycopene, a naturally occurring constituent of tomato, is a nutritional factor related to Vitamin A. Tomato varieties with transgenically enhanced lycopene content are under investigation. Another trait of interest is delayed ripening. Tomatoes that ripen slower can remain on the vine longer and develop improved flavour, compared to commercial varieties that are picked at the green stage.

The Flavr-Savr® tomato, one of the earliest approved transgenic crop varieties, was a delayed ripening variety. Because the trait was incorporated into a variety that performed poorly otherwise, it was not a commercial success. Salty soils are an increasing problem in many parts of the world. Many crop plants, including tomatoes, are killed by high salt levels in soil and irrigation water. The development of a salt-tolerant tomato offers the possibility that tomatoes could be grown on land that was previously unavailable for agriculture.

Golden rice

Millions of people in the world suffer from Vitamin A deficiency, which leads to vision impairment and increased susceptibility to diarrhea, respiratory diseases, and measles. Rice is a staple food in many countries, particularly in Asia, but does not contain Vitamin A or its immediate precursors. By inserting two genes from daffodil and one gene from a bacterial species into rice plants, Swiss researchers have produced rice capable of synthesising beta-carotene, the precursor of Vitamin A. This rice variety is now being crossed into adapted varieties, with field tests possible in a year or two.

Canola

Canola is a major oilseed crop. Transgenic research has focused on improving the nutritional quality of canola oil by enhancing the Vitamin E content or by modifying the balance of fatty acids.

Turfgrass

There are a number of environmental concerns associated with the current use of turfgrass. These include

— the amount of chemicals applied (fertilizers, herbicides, fungicides, and even green dyes);

— the large quantity of water required to keep lawns healthy, especially in the West;

— the energy required to mow lawns.

New transgenic turfgrass varieties will address some of these concerns by incorporating genes for herbicide tolerance, disease and insect resistance, reduced growth rates (meaning less mowing), and tolerance to drought, heat, and cold. The first such product on the market will most likely be Roundup Ready® varieties of Kentucky bluegrass, creeping bentgrass, and buffalograss.

Plant-based vaccines

Food crops engineered to produce edible vaccines against infectious diseases would make vaccination more readily available to children around the world. Because of their palatability and adaptation to tropical and subtropical environments, bananas have received considerable research attention as a vehicle for vaccine delivery. Transgenic bananas containing inactivated viruses that cause cholera, hepatitis B, and diarrhea have been produced and are currently undergoing evaluation.

Sunflower: A disease-resistance trait, an anti-pest trait, and a herbicide-resistance trait are all being pursued, but no commercial varieties are now on the market. White mold (Sclerotinia) is a serious problem for sunflower producers in some areas. Resistance to this disease would expand the area in which sunflowers can be grown and might improve yields in areas of current cultivation. Resistance to the Argentina looper, an insect that eats sunflower leaves, is also being investigated. Some research has been done on developing sunflowers that can tolerate being sprayed with the weed killer Roundup. This trait would allow a farmer to spray his field for weeds while not killing the crop.

Coffee and Tea: Decaffeinated coffee is now made by treating coffee beans to remove the caffeine. One method uses organic solvents to extract the caffeine, which causes some consumers to be concerned that residues from the solvents will remain in the coffee they drink. Other methods are criticised for removing some of the desirable, flavour-producing components along with the undesirable caffeine. The harvesting of coffee beans now requires many passes through the orchards because the beans ripen at different times. A scientist in Hawaii is developing a method for making all the beans ripen at the same time so that harvesters could pick all the beans during one pass through the orchard.

Grapes and wine: Grape vines (Vitis vinifera) are susceptible to several diseases that reduce the amount and the quality of wine grapes and table grapes or even kill the vine. Genes that confer resistance to particular diseases would reduce the cost of battling diseases in the vineyard.

Tobacco: Nicotine-free tobacco is now being grown for a projected introduction of nicotine-free cigarettes. Previous attempts to make low-nicotine products removed some of the flavour along with the nicotine. Genetically engineered nicotine-free tobacco doesn't synthesise nicotine in the leaf.

Trees: Forest trees such as poplar, aspen, and spruce have been transformed with various genes to provide resistance to insects, tolerance to herbicides, and higher levels of the commercial product. For example, reducing the lignin content of a tree can make it easier to recover wood pulp.

6

Biotechnology for Plant Micropropagation

All plant somatic cells once isolated and cultured in vitro are capable of expressing totipotency. The injured cells in the outer layers of the isolated explant evolve ethylene that induces the inner layers of cells to undergo dedifferentiation. After the loss of coordinated control, which is ensued by cell division, the formation of new gradients of endogenous phytohormones in the dedifferentiated cells enhance cell divisions in response to the various growth regulators added to the medium.

Cell division can take place in an unorganised pattern with callus formation, or in an organised pattern with the formation of meristematic centers directly in the explant tissues. Redifferentiation in the callus tissue takes place when new physiological gradients are formed in the non-organised parenchymatous tissue along with the formation of meristemoids (in vitro meristems), which can further differentiate into organised structures. In several plant species, meristematic centers form directly on the explant with very little or no callus formation and can develop into either shoots or somatic embryos.

Plant regeneration can follow after the initial stages of controlled redifferentiation through either the organogenic or embryogenic pathways. Plant regeneration in vitro is dependent on the manipulation of the inorganic and organic constituents in the medium, as well as the type of explant and the species. In most plants, successful regeneration from the callus or directly from the explants takes place after a series of subcultures in various media, in a sequence which is often specific to the species, variety, or the newly introduced genotype. The determining factors are the combination of the concentration in relation to mediumvolume and the composition of growth promoting and retarding regulators in the medium, the physiological status and competence of the cells and their capability for morphogenetic expression.

Micropropagation (in vitro propagation of axillary and/or adventitious buds as well as somatic embryos) is presently used as an advanced biotechnological system for the production of identical pathogen-free plants for agriculture and forestry. The technique, however, is still costly due to intensive hand manipulation of the various culture phases and is not used commercially for all plant species. In addition, in some plants, the initial stage of establishment and response is slow and the survival of the plants in the final stage ex vitro is often poor, which further reduces the micropropagation production potential. Efficient commercial micropropagation depends on rapid and extensive proliferation along with the use of large-scale cultures for the multiplication phase.

Furthermore, normal plant development during the acclimatisation and hardening stage is mandatory to ensure a high percent of survival after transplanting to the greenhouse. Mechanisation and automation of the micropropagation process can greatly contribute to overcoming the limitation imposed by existing conventional labor-intensive methods. Considerable attention has been directed toward automation of the repeated cutting, separation, subculture, and transfer of buds, shoots, or plantlets during the multiplication and transplanting phases.

Progress in tissue culture automation will depend on the use of liquid cultures in bioreactors, which will allow fast proliferation, mechanised cutting, separation, and automated dispensing. These techniques were reported for some plants and were shown to reduce hand manipulation and thus reduce in vitro plant production costs. Liquid media have been used for plant cells, somatic embryos, and organ cultures in both agitated flasks or various types of bioreactors. Although the use of bioreactors has been directed mainly for cell suspension cultures and secondary metabolites production, research directed at improving bioreactors for somatic embryogenesis has been reported for several plant species.

The cultivation in liquid media using a temporary immersion system with different frequencies of immersion was reported to improve plant quality and multiplication rates of banana, coffee, and rubber. Bioreactors were used also for the cultivation of hairy roots mainly as a system for secondary metabolite production. The use of a simplified acoustic window mist bioreactor for transformed roots and for carnation shoots was reported to improve biomass growth. Information on the use of bioreactors as a system for plant propagation (Table 1) through the organogenic pathway, although limited to a small number of plant species, is presently being applied to several ornamental and some vegetable and fruit crop plants.

Table 1. Plants propagated in bioreactor cultures.

Species	Response
Amaryllis hippeastrum	Buds, plants, bulblets
Ananas comosus	Shoot clusters
Apium graveolens	Somatic embryos
Araceae species	Shoots, plants
Brodiaea complex	Bud clusters, corms
Coffea arabica	Shoot clusters, plants
Cyclamen persicum	Callus, somatic embryos
Daucus carota	Callus, somatic embryo
Dianthus caryophyllus	Shoots, plants
Eschcholtzia californica	Somatic embryos
Euphorbia pulcherrima	Somatic embryos
Fragaria ananasa	Shoots, plants

Gladiolus grandiflorum	Bud clusters, plants, corms Shoots, plants, corms
Hevea brasiliensis	Buds, plants
Hyacinthus orientalis	Bulblets, plants
Lilium spp.	Plants, bulblets
	Bulblets, plants
Medicago sativa	Callus, somatic embryos
Musa spp.	Buds, plants
Nephrolepis exaltata	Buds, plants
Nerine sarniensis	Proembryogenic clusters, somatic embryos, bulblets
Ornithogalum dubium	Shoot clusters, plants, bulblets
Populus tremula	Bud clusters, shoots, plants
Picea glauca	Somatic embryos
Picea glaucaengelmannii	Somatic embryos
Picea marianna	Somatic embryos
Solanum tuberosum	Plants, tubers Bud clusters, plants, tubers Shoots, plants, tubers

However, liquid cultures confer several problems associated with abnormal plant development, a phenomenon described as hyperhydricity, previously termed "vitrification", which causes poor plant development in vitro and later in ex vitro. Attempts were carried out to overcome hyperhydricity in liquid cultures by the use of growth retardants, inhibitors of gibberellin biosynthesis, which decreased hyperhydricity, reduced shoot elongation, and induced bud or meristem cluster formation. The clusters were shown to be an alternative propagation system for bioreactor cultures, providing a biomass with limited leaf elongation. Recently micropropagation of pineapple was scaled up using clusters in a temporary immersion system (ebb and flow) in bioreactors.

Developmental Pathways of Plant

Somatic Embryogenesis

The use of liquid cultures for the cultivation of somatic cells in recapitulating embryogeny was reported by Steward et al. and Reinert as early as 1958 in carrot (Daucus carrota L.). The underlying concept for cell totipotency and the ability to renew growth and express morphogenesis was based on the assumption that isolation and bathing of the cells in a specific medium will simulate the zygotic embryo conditions in the ovary. The bathing liquid medium that contains various nutrients and growth regulators was assumed to be a source for similar stimuli that were present in the zygotic embryos' immediate environment.

In addition, the segregation of embryogenic from non-embryogenic cells amplified the expression of cell totipotency. Somatic embryogenesis was achieved by the initial use of an induction medium containing an auxin, usually 2,4-dichlorophenoxyacetic acid (2,4-D), and coconut water; the latter was mainly a source of cytokinins, inositol, and reduced nitrogen compounds. Once the pro-embryonic clusters formed, the removal or lowering of the auxin concentration initiated a sequence of events similar to zygotic embryogeny. Somatic embryogenesis was observed at first mainly in members of the Umbelliferae cultured in liquid media and has since been reported in gymnosperms and angiosperms, including several ornamentals, and vegetable and field crops, as well as perennial woody plants.

The embryogenic pathway, as opposed to the organogenic pathway, can be a more efficient and productive system for large-scale clonal propagation. Somatic embryogenesis may result in less variation through chimerism. However, somaclonal variation will be less likely to take place in direct regeneration, but might be more pronounced in somatic embryos developing from continuously cultured callus tissue. Since the embryo contains both a root and an apical shoot meristem, the rooting stage required in conventional in vitro bud or shoot propagation technology is obviated.

Somatic embryos are small and can be adequately handled in scaled-up procedures. They are amenable to sorting and separation by image analysis, dispensing by automated systems, and can be encapsulated and either stored or planted directly, with the aid of mechanised systems. Somatic embryogenesis in liquid shake or bioreactor cultures was reported in carrot, in caraway, poinsettia, alfalfa, celery, in Eschcholtzia californica, in Nerine, in Octea catharinensis, in sweet potato, in rubber, and in spruce. However, the ultimate goal of production of synthetic seeds through successful encapsulation of somatic embryos from liquid shake or bioreactor cultures was reported only for carrot, alfalfa, celery, and white spruce and is currently under continuous investigation for several other species.

Organogenic Pathway

The propagation of most plants is presently carried out commercially through the organogenic pathway in agar-gelled cultures, even though the protocols are long and costly. The advantages for the mechanisation of the process are mainly for achieving a reduction in hand manipulation and labor costs. The process has to be scaled up using liquid cultures in bioreactors to amend it to automation. The information on the use of bioreactors for unipolar structures such as protocorms, buds, or shoots is limited, mainly because of the problems of hyperhydricity of the leaves and shoots.

One of the first attempts in using liquid cultures for micropropagation of buds was reported for orchids that formed protocorms with minimal shoot elongation during the liquid culture stage. The protocorms could be separated and induced to form new plants after subculture to agar-gelled medium. Several other ornamental plants also have been propagated through the organogenic pathway in liquid shake cultures and bioreactors. By controlling shoot growth and providing culture

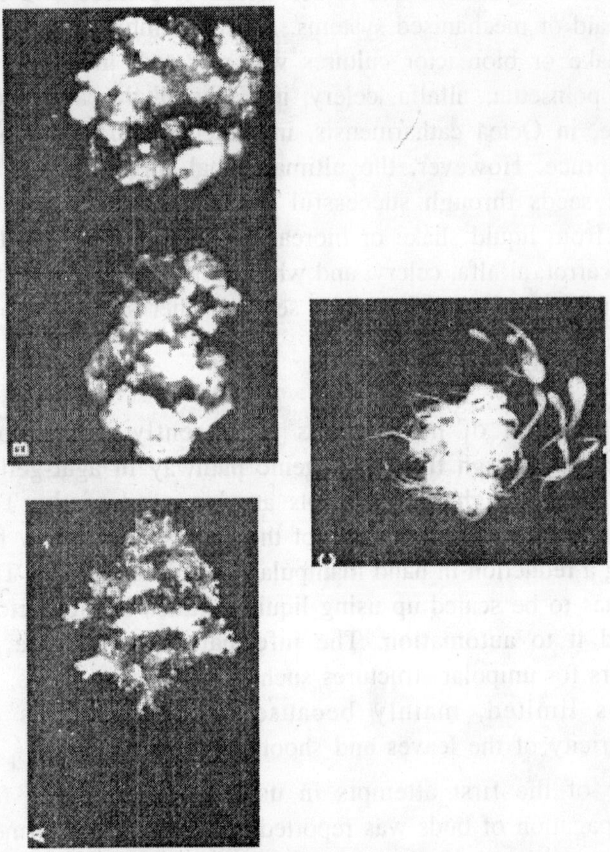

Figure 1. Liquid cultured bud clusters of potato (A), meristematic clusters of gladiolus (B), and bulblet formation in Ornithogalum dubium bud clusters (C).

conditions that reduced abnormal leaf growth and enhanced the formation of bud or meristematic clusters, as was shown for potato, gladiolus, and Ornithogalum dubium (Fig. 1A, B, C), a high proliferation rate was achieved without the phenomenon of hyperhydricity.

Meristem Clusters

The development of spherical meristematic or bud clusters in liquid cultures was shown to provide a highly proliferative and rapid growing system amenable to automated inoculation, control of the medium components, mechanical separation, and efficient delivery to the final stage for plant growth and development. Cluster formation appears to be associated in most species with the continuous submergence, circulation, and agitation of the plant biomass in the medium, as well as with a balanced ratio of growth-promoting and growth-retarding regulators. Clusters can form from axillary buds and/or adventitious buds as well as from meristemoids in pro-embryogenic callus that later differentiate to somatic embryos (Fig. 1A, 2A, B).

The formation of condensed organised structures in which the shoots are reduced to buds or meristematic tissue was reported in several plant species. The clusters were made up of densely packed meristematic cells, actively dividing and forming new meristematic centers on the outer surface. The meristemoids surrounded loosely packed cells in the center and exhibited some vascularisation, as was shown in liquidcultured poplar clusters. In banana, on the other hand, the clusters were made up of condensed buds surrounding a central core with a cavity.

Halperin in 1967 described the formation of pro-embryogenic masses in carrot root explants which were induced by the addition of 2,4-D to the liquid medium, resembling orchids protocorm-like bodies that formed when isolated buds were cultured in liquid medium in shake cultures. Protocorm-like clusters also were induced in liquid-cultured gladiolus buds and in several species of the complex Brodiaea by the addition of paclobutrazol or ancymidol (gibberellin biosynthesis inhibitors) to the medium. In chicory, nodules were induced on leaves by IBA and BA in the medium.

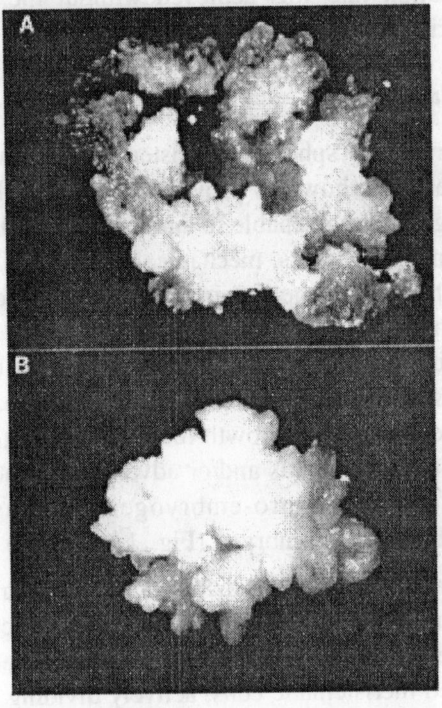

Figure 2. Nerine proembryogenic clusters in a 2,4-D medium (A) and after 2,4-D removal with the addition of 2iP showing somatic embryos (B).

In potato and banana, bud clusters were induced by a balanced ratio between kinetin and ancymidol in the liquid medium. In woody species, nodules in poplar in liquid medium and in radiata-pine in agar-gelled medium respectively, induced by the use of a balanced ratio of growth regulators. Levin et al., working with several ornamental species, described a severalfold increase of an organogenic biomass of clusters which proliferated in bioreactors and were separated mechanically prior to dispensing to agar-gelled cultures for further growth.

The production of clusters in Philodendron cultured in liquid medium required the presence of benzylaminopurine (BA) and an inductive treatment for 24-48 h. with ancymidol. The inductive treatment prevented a carry-over dwarfing effect of ancymidol on leaf and shoot development after transplanting the clusters to agar-gelled medium for further plant growth. Gladiolus clusters cultured in a disposable 2-L bioreactor proliferated and made up to 60% of the vessel volume after 4-5 weeks. The aggregates which reached 0.5-1.5 cm in diameter tended to sink to the bottom of the bioreactor as the biomass increased and could not be resuspended unless the rate of aeration was increased.

An increase from 0.5 to 1.5 vvm (volume air/volume medium per minute) provided the required aeration for recirculation of the cluster biomass. Potato and banana bud clusters proliferated in bubble and dis- posable plastic bioreactors at a faster rate than on agar-solidified medium. Growth regulators, when given in a specific balanced ratio of promoting and retarding substances, apparently act as morphogenic signals and control the development of the spherical meristematic or bud clusters.

The spherical meristematic or bud clusters were described by various terms: pro-embryogenic mass (PEM) in carrot, protocorms in orchids, nodules in poplar and radiata-pine, nubbins in daylilly, and meristematic or bud clusters in gladiolus, Brodiaea, Nerine sarniensis, fern, Philodendron, potato, banana, and Ornithogalum dubium. It is suggested that the term cluster should be used, as it is more appropriate for describing these proliferative, usually rounded structures, which develop from buds or other proliferative tissue in liquid culture, through either the organogenic or somatic embryogenesis pathways.

Anomalous Plant Morphogenesis

Scaling-up in bioreactors requires the use of liquid instead of agar-gelled media during the proliferation and biomass production stages. The culture of plants in liquid medium is known to cause anomalous morphogenesis, resulting in plant hyperhydricity. The plants that develop in liquid media are fragile, have a glassy appearance, with succulent leaves or shoots and a poor root system. The leaves are the organs affected most severely in liquid cultures. They develop an unorganised mesophyll tissue that is made up mainly of spongy parenchyma tissue with large intercellular spaces, a deformed vascular tissue, and an abnormal epidermis.

The epidermal tissue in hyperhydric leaves lacks a welldeveloped cuticle and possesses malfunctioning guard cells which cannot respond to closure signals. Hyperhydricity affects plant survival after transplanting and causes loss of the in vitro developed leaves, or even whole plants, which often wilt and die. The two major processes carried out by the leaves, photosynthesis and transpiration, are not fully functional in hyperhydric leaves and thus cause the poor performance of the transplanted plants ex vitro. In many plants propagated in vitro, the anomalous morphology and anatomical aberration often observed in plants result from deviation from the normal course of morphogenic events that are manifested in plants in vivo. Some of the deformation, such as hyperhydric, malformed leaves and shoots, abnormal embryos, recurrent embryogenesis, and several other disorders, are apparently the result of interruption or faulty timing of the signals involved in the normal sequence of organisational events known to exist in vivo.

Cell and Tissue Growth in Bioreactors

The advantages provided by aerated liquid cultures in bioreactors include better contact between the plant biomass and the medium, no restrictions of gas exchange, the control of the composition of both the medium and the gaseous atmosphere, and the ability to manipulate the plant biomass in relation to the medium volume. The need for efficient circulation and mixing of the plant biomass, especially for cluster and embryogenic tissue, is essential to prevent sedimentation and allow optimal growth. Various types of bioreactors with mechanical or gas-sparged mixing were used for plant cell cultures, to provide stirring, circulation, and aeration.

Mechanically stirred bioreactors depend on impellers, including a helical ribbon impeller, magnetic stirrers, or vibrating perforated plates. Aeration, mixing, and circulation in bubble column or airlift bioreactors is provided by air entering the vessel from a side or bottom opening through a sparger; as the air bubbles rise, they lift the plant biomass and provide the oxygen gas required. Mechanically stirred bioreactors are used for effective mixing, aeration, dispersion of air bubbles, and prevention of large cell aggregates formation in cell suspension cultures.

In recent years, large-scale cultivation of plant cells, embryos, or organs has made use of airlift or bubble column bioreactors and, to a lesser extent, of mechanically stirred tank bioreactors, due to the former lower shearing force properties. Plant cells are relatively large in size, have sensitive cell walls which are highly susceptible to shearing forces, and are easily damaged. Furthermore, since plant cells, unlike microbial cultures, do not require high O_2, the use of bubble column or airlift bioreactors was found to be adequate and advantageous for plant tissue cultured in liquid media.

In addition, it was shown that mixing by gas sparging in bubble column or airlift bioreactors lacking impellers or blades is far less damaging for clusters than mechanical stirring, since they were shown to have a lower shearing stress. The main advantage of airlift bioreactors is their relatively simple construction, the lack of regions of high shear, reasonably high mass and heat transfer, and relatively high yields at low input rates. A bubble free oxygen supply bioreactor with silicone tubing was found suitable for embryogenic cell suspensions and provided foam-free cultures.

For hairy root culture, an acoustic mist bioreactor was found to increase root biomass significantly. For efficient large-scale cultures of both somatic embryos and organogenic plant tissue, the bioreactor configuration and volume must be determined according to the mixing and aeration requirements of the specific plant or tissue propagated, as well as for minimising the intensity of the shear stress. One major problem encountered in large-scale liquid cultures is contamination. Fungi, bacteria, yeast, and insects are a source of serious contaminations, causing heavy losses of plant material in commercial laboratories.

In scaled-up liquid cultures, the losses are even greater, as the source of contamination due to manipulation of the bioreactor apparatus is dependent on the various stages of preparing and maintaining the equipment. In several laboratories, the operation area is kept sterile by a positive pressure airflow, which decreases contamination risks. In addition, constant screening of the plant tissue for contaminants and continuous indexing is another safeguard.

Factors in Liquid Cultures

In order to control plant morphogenesis and biomass growth in bioreactors, various culture conditions must be manipulated, i.e., the gaseous atmosphere, oxygen supply and CO_2 exchange, pH, minerals, carbohydrates, growth regulators, and the liquid medium rheology and cell density.

Gaseous Atmosphere

The atmosphere of the culture vessel is made up mainly of nitrogen (78%), oxygen (21%), and carbon dioxide (0.036%). The culture vessel gas composition is influenced by the volume of the vessel and the extent of ventilation. Plants evolve CO_2 and consume O_2 during respiration, while during photosynthesis CO_2 is used and O_2 is produced. During the dark period, CO_2 levels were found to increase in cultures, and if photoautotrophic conditions prevail, its level decreased in the light. Ethylene, ethanol, acetaldehyde, and other hydrocarbons are additional components of the gaseous atmosphere in vitro.

Most of the effects of CO_2, O_2, and C_2H_4 on plant growth in vitro were reported for agar-gelled or cell suspension cultures. In bioreactors, the control of the gaseous phase depends on the gas flow and can be easily manipulated to provide the required levels of O_2, CO_2, and C_2H_4. In airlift or bubble column bioreactors, the air supplied is used for both mixing and aeration. The importance of aeration and the gaseous phase were shown in potato cultured in airlift bioreactors. Induction of tubers was inhibited under continuously submerged conditions.

Microtubers developed only after the shoots elongated and reached the gaseous phase at the top of the bioreactor. Enrichment with O_2 and manipulation of the hormonal and osmotic conditions had no effect and could not explain the phenomenon. A two-phase culture substituting the growth medium with a tuber induction with 9% sucrose in the medium enhanced tuber formation from shoots that developed above the medium and were exposed to the gaseous phase.These results emphasise the importance of the gaseous phase in the bioreactors for specific developmental phases.

Oxygen Level: Oxygen levels in liquid cultures depend on the presence of O_2 in the gas phase above and in the air bubbles inside the medium, as well as in the dissolved O_2 in the medium. Air is released through a sparger located at the base of the bioreactor. The available oxygen for plant cells in liquid cultures determined by oxygen transfer coefficient (k_La values) is the part that dissolves in water. Its depletion as a function of the metabolic activity of the growing cell biomass can affect the culture yield. Plant cells have a lower metabolic rate than microbial cells and a slow doubling time and therefore require a lower O_2 supply. In general, high aeration rates appear to reduce the biomass growth.

The requirements for O_2 may vary from one species to another, and must be supplied continuously to provide adequate aeration, since it affects metabolic activity and energy supply as well as anaerobic conditions. The level of O_2 in liquid cultures in bioreactors can be regulated by agitation or stirring and through aeration, gas flow, and air bubble size. The use of a porous irrigation tube as a sparger generated fine bubbles, high kLa values, low mechanical stress and provided a high growth rate. Growth of poinsettia cell suspension in bioreactors was inhibited when the level of O_2 dropped below 10%.

When the level of O_2 was elevated to 80%, cell number increased to 4.9×10^5/ml as compared to 3.1×10^5/ml at 40% O_2. Somatic embryo development in alfalfa and poinsettia suspension cultures was enhanced at 78% and 60% O_2 levels, respectively. High aeration rates were found to inhibit cell growth in cell suspensions cultured in airlift bioreactors. This result was explained to be due to an effect of "stripping" of the volatiles produced by the plant cells, which are apparently necessary for cell growth.

Increasing O_2 levels from 21% to 80%, in bioreactor cultures of Boston fern clusters, enhanced growth values (final FW -initial FW/initial FW) from 0.61 to 0.92. Reducing O_2 levels to 10% (v/v) affected cell differentiation in bioreactor cultures of carrot embryogenic tissue. Under these conditions embryo production was severely inhibited. The best production of somatic embryos from embryogenic cultures of Eschcholtzia californica and carrot in a helical ribbon impeller bioreactor was achieved under 20% oxygen concentration. When O_2 was low (5-10%), it inhibited biomass and somatic embryo production, while high O_2 (60%) favoured undifferentiated biomass production. Additional studies are required to provide information on optimal dissolved O_2 requirements in large-scale liquid cultures.

CO_2 Effects: The effects of CO_2 were reported mainly for plants in agargelled cultures or for cell suspension cultures used for secondary metabolite production. The reports on the effects of CO_2 enrichment in sugar-free agargelled cultures suggest beneficial promotion of plant growth during plant acclimatisation and transplanting ex vivo. The contribution of CO_2 supply during the proliferation and multiplication stage in media supplied with sucrose in bioreactors is debatable. It is implicit that if

photoautotrophic conditions do not prevail, CO_2 enrichment beyond the 0.36% in the air supply is unnecessary.

There are reports that high aeration rates rather than excessive oxygen inhibit growth and that reduced growth could be due to depletion of CO_2 or to the removal of various culture volatiles including CO_2. The requirement for CO_2 was not related to photosynthesis but to some other metabolic pathways involved in amino acid biosynthesis. In poinsettia, the growth of embryos in Erlenmeyer flasks was considerably higher than in bioreactors and it was suggested that the growth difference was related to the differences in the gaseous atmosphere in the headspace. CO_2 enrichment in an illuminated bioreactor culture of Brodiaea clusters did not affect biomass growth. An increase from 0.3% to 1% gave a similar growth value under the two CO_2 levels and supply of 135 μmols · m^{-2} · s^{-1} photosynthetic photon-flux. In Cyclamen persicum Mill, high CO_2 levels correlated with increased production of pro-embryogenic masses.

Ethylene: Ethylene level in the headspace in liquid cultures in flasks differs from that in continuously aerated bioreactor cultures. Most of the reports on ethylene relate to agar-gelled or cell suspension cultures. In various plant species cultured in bioreactors, the rate of aeration will affect the level of ethylene. High rates of aeration, which are often required at high biomass densities, can cause "stripping" of volatiles that are apparently important for some plants grown in culture. In clusters of Brodiaea cultured in liquid medium, ethylene had no effect on growth, although its level was reduced in the presence of silver thiosulfate (an inhibitor of ethylene action). The level of ethylene was reduced from 0.38 to 0.12 ml/L in highly aerated bioreactor cultures without affecting biomass growth, which was similar under both levels.

Consumption of Mineral Nutrients

Revised MS medium or media with various partial modifications in the inorganic and organic constituents of MS are used for most plant species in agar-gelled or liquid cultures in vitro. The availability of mineral nutrients depends on the type of culture, whether agar-gelled or liquid, the type and size of the plant biomass, and the physical properties of the culture. Factors such as pH, temperature, light, aeration, the concentration of minerals, the medium volume, and the viscosity of the medium will determine the rate of absorption of the various nutritional constituents. Plant cells growing in liquid cultures are better exposed to the medium components and the uptake and consumption are faster.

Agar-gelled and liquid shaken cultures often dehydrate by water evaporation to the headspace and out of the vessel and such a water loss concentrates the medium. The content of a certain component in the medium is a product of its concentration and the medium volume. In bioreactors, in which either humidified air or condensers are used to prevent dehydration, the level of the nutrients in the medium is affected mainly by the absorption rate and by cell lysis.

Differentiation and proliferation of micropropagated fern, gladiolus, and Nerine nodular clusters in bioreactors was better on half-strength than on full-strength MS minerals. This was also true for Lilium bulblets differentiating on bulb scales that were cultured in bioreactors. A drop in pH to 4.5 and lower values and the subsequent increase to pH 5.5 was attributed to the initial utilisation of ammonium and to the uptake at a later stage of nitrate. In several species the depletion of NH^+_4 is the first limiting factor of biomass growth and somatic embryos development.

Increasing the concentration of NH^+_4 to 15 mM resulted in maximum somatic embryo production in carrot cultures in a helical ribbon impeller bioreactor. A thorough and detailed study of nutrient uptake in Eschcholtzia californica embryonic cultures was achieved in a helical ribbon impeller bioreactor. Following a lag phase of about 140 h, first NH^+_4 and then NO^-_3, PO^-_4, K^+ and SO^-_4 uptake was observed to coincide with biomass increase and somatic embryo production. After 300 h, PO^-_4 was exhausted, while after 600 h NH^+_4 was depleted and Mg^{++} and Ca^{++}

composition continued to deplete after about 800 h. The K^+ and Na^+ ions were not depleted at that stage.

The composition of minerals, monitored during the culture of Begonia and Brodiaea in liquid media changed, with phosphate, ammonia, nitrate, and potassium depleting faster and prior to the depletion of Ca^{++} and Mg^{++}. In somatic embryos of spruce cultured in bioreactors, 80% of the ammonium was consumed by the growing biomass. In general, biomass growth is limited by the availability of phosphate, nitrogen, and carbohydrates and to a lesser extent by the availability of calcium, magnesium, and other ions.

Supply of Carbohydrate

Cultured plants require a constant supply of carbohydrates as their source of energy. Sucrose and to a lesser extent glucose, fructose, or sorbitol are the most commonly used carbohydrates in vitro. In general, sucrose is removed rather rapidly from the medium and after 10-15 days the sucrose can be completely depleted or reduced to 5-10 g/L from an initial level of 30 g/L in both agar-gelled and liquid cultures. At the same time, glucose and fructose that appear in the medium due to sucrose hydrolysis increase in the presence of invertase in the culture medium, and can reach levels of 5-10 g/L.

Cell suspensions of Catharanthus roseus cultured in a column airlift bioreactor showed a lag phase of 5 days, during which there was a total hydrolysis of the sucrose to glucose and fructose. In suspension cultures of alfalfa, sucrose also was hydrolyzed during the first 5 days. Most of the sugars uptake occurs after day 5 and glucose is taken up preferentially over fructose. A higher yield of alfalfa embryos was obtained when 30 g/L maltose combined with NH^+_4 was used instead of 30 g/L sucrose that was used in combination with various nitrogen sources. In embryonic suspension cultures of celery, the addition of mannitol reduced cell lysis and enhanced somatic embryogenesis.

When 40 g/L mannitol was added, a higher number of embryos was produced and the frequency of singulated normal embryos was increased. In embryogenic cultures of Eschcholtzia californica grown in a helical ribbon impeller bioreactor, sucrose uptake started after 100 h and the sucrose was depleted after 600-800 h in culture. In Picea species cultured in shaken flasks in a medium with 30 mM glucose, the carbohydrates depleted after 10-14 days. In mechanically stirred bioreactors, a level of 60 mM sucrose resulted in the highest cell biomass and somatic embryo number.

The effects of various carbohydrates on the growth of spruce somatic embryos revealed that the response was species dependent. The biomass of fern meristematic clusters in a bubble column bioreactor was increased with the increase in sucrose concentrations from 7.5 g/L to 30 g/L, while higher levels caused a decrease in cluster growth. Elevated sucrose concentrations caused a decrease also in the chlorophyll content of the clusters and leaves. Increasing sucrose concentrations from 30 g/L to 60 g/L in bioreactor cultured gladiolus clusters decreased the biomass FW by more than 50%. When 30 g/L or 60 g/L were combined with paclobutrazol, a further decrease in FW was observed.

On the other hand, 60 g/L sucrose induced a higher DW increment, which could be attributed to an osmotic effect and the subsequent water status of the clusters. Gladiolus clusters cultured in the presence of growth retardants had a higher level of starch, 845 as compared to 585 mg/g DW in the control. A higher number of bulblets were produced in bioreactor-cultured bulb scales of Lilium at 30 g/L than at higher sucrose levels. However, larger size microbulbs were produced at 90 g/L than at 30 g/L sucrose in the medium.

Effects of pH

The initial pH in most plant cell cultures ranges between 5.5-5.9. Since most media are not buffered, changes during autoclaving and during the biomass growth in culture occurs. A rapid drop in pH to 4.0-4.5 took place within 24-48 h in cell suspension, organ, and embryogenic cultures. These changes were related to an initial ammonium uptake and acidification due to cell lysis. However, the pH increased after a few days and reached a stable level around pH 5.0-5.5, which was related to the uptake of nitrates. In spruce species cultured in liquid medium, the pH levels were shown to increase to 6.5-6.8 after 14 days in culture.

In embryogenic cultures of poinsettia, various initial pH levels stabilised after 20 days in batch bioreactor cultures, between pH 5.3-5.7. Oxygen levels affected pH changes by 0.2 pH units. In fern cultured in an airlift bubble column bioreactor, the pH dropped to 4.2 after 24 h and subsequently increased to pH 5.4. Keeping a constant pH by titration with KOH did not affect the growth response. In alfalfa, the development of somatic embryos was affected by the pH. A higher rate of embryo production was observed at a constant pH of 5.5 than at a nonbuffered medium or at lower pH levels.

Carrot cell differentiation in a controlled bioreactor was affected by the pH; the highest rate of embryo production was observed at a pH of 4.3. However, embryo development was arrested before the embryos reached the torpedo stage and continued only at pH 5.8. The changes in carrot embryo development were associated with sugar uptake and ammonium depletion and can be attributed to enzyme and metabolic activity at an optimal pH. It appears that pH requirements are species and developmental stage dependent.

Effects of Growth Regulator

The use of growth regulators in liquid cultures can be more effective in controlling the proliferation and regeneration potential than in agargelled medium due to the direct contact of plant cells and aggregates with the medium. The limited information available on regulators seems to indicate that similar levels of growth regulators were used in both agar-gelled and liquid cultures even though availability appears to be better in liquid medium. Somatic embryogenesis in many species was first induced in an auxincontaining medium that promoted rapid cell division.

Expression of the embryogenic potential was achieved in auxin-free media, or with low levels of auxin in the medium. Further embryo conversion and maturation into plants was promoted by the addition of abscisic acid (ABA), which was found to control abnormal embryo and plant growth. Bulblets development from scales in Lilium cultured in bioreactors was higher in the presence of BA than kinetin. However, high BA had an inhibiting effect on further bulblet growth. High kinetin also stimulated bulblets differentiation, but inhibited further growth of the bulblets.

However, the addition of 0.1 mg/L naphthalenacetic acid (NAA) was found to further enhance kinetin activity, which was, however, inhibited again in the presence of high (90 g/L) sucrose levels. In embryogenic cultures of Nerinė, auxin and cytokinins were used to induce proembryogenic clusters. Embryogenic expression was achieved, however, only after a short exposure to 2-iso-pentenyladenine (2iP) and further subculture in a growth regulator-free medium. Since one of the major problems in liquid-cultured plants is malformation of shoots, hyperhydricity, the induction of meristematic or bud clusters with arrested leaf growth was one of the solutions to hyperhydricity.

The use of relatively high cytokinin levels and growth retardants which inhibit gibberellin biosynthesis was the most effective method to reduce shoot and leaf growth and promote the formation of meristematic clusters. The information on abscisic acid is limited and was found effectual mainly in the later stages of somatic embryo development, promoting normal embryo growth and maturation in carrot and alfalfa embryogenic cultures. Growth of callus and leaf development in bulblets regenerated on bulb scales of Lilium in bioreactor cultures was inhibited by abscisic acid, thus providing singulated propagules for easier handling and storage. Ethylene effects appear to be species dependent.

Effects Temperature

The control of the temperature in the liquid medium inside the bioreactor can be easily manipulated by a heating element in the vessel or by circulating water in an enveloping jacket outside the vessel. There is, however, limited information on the effects of temperature in bioreactor cultures, which is usually kept constant at 25°C, with small day and night fluctuations. Temperature effects on potato tuber formation in an airlift bioreactor were studied by Akita and Takayama (1994). A higher number and larger-size tubers developed at 25°C than at 17°C; the lower temperature caused a decrease in tuber size.

Cell and Aggregate Density

Growth and proliferation of the biomass in bioreactors depends on airflow supply for the aeration and mixing, and for the prevention of the plant biomass sedimentation. In many plants cultivated in bioreactors, continuous aeration, mixing, and circulation cause shearing damage, cell wall breakdown, and accumulation of cell debris, which is made up mainly of polysaccharides. Cell debris accumulation results in foaming, adhesion of cells and aggregates to the culture vessel walls, and the development of a "crust" at the upper part of the bioreactor vessel. This layer prevents adequate circulation, causing additional cell debris formation and a demand for higher aeration rates that intensify the clogging problem.

As the biomass increases and the cultures become viscous, higher rates of aeration are required to allow for oxygen supply and circulation. Medium viscosity and foaming were reduced by the use of half the concentration of MS minerals and by lowering the level of calcium in the medium. In bioreactor cultures in which the plant biomass used was of an aggregate or cluster nature, the foaming and viscosity were not as severe as that observed in cell suspension cultures. Surprisingly, in cluster cultures in bioreactors, even when the mixing was poor and the clusters sedimented, biomass growth was observed, suggesting that perhaps the continuous circulation accompanied by aeration was secondary in importance to oxygen supply.

Shearing stress and cell wall damage were greatly reduced in a Vibramix bubble-free oxygenated bioreactor, in which silicone tubing was used for air supply. Embryogenic cells cultured in a helical ribbon impeller bioreactor produced poor-quality embryos when the mixing speed was increased from 60 to 100 rpm. This was attributed to shearing stress and cell damage. The introduction of polyethylene glycol 6000 changed the rheology of the medium in alfalfa liquid cultures and improved somatic embryo development beyond the globular stage while it was arrested in a less viscous medium.

Disposable plastic presterilised bioreactors, with a volume capacity of 2 and 5 L, were used for the micropropagation of various plant species through bud or meristematic clusters. The plastic bioreactors were found to provide good circulation with reduced shearing damage and foaming. The problems of cell damage, foaming, and culture viscosity can be better controlled by developing bioreactors with an optimal shape suitable for micropropagation through meristematic or bud clusters. Clusters that are compact and consist mainly of small meristematic cells are apparently less shear sensitive than highly vacuolated cells in parenchymatous cell suspension cultures.

7

Biotechnology for Improve Orphan Crops

Orphan crops are not produced widely around the world, they are not traded to any significant extent in international markets, and they receive considerably less attention than the major crops from international or regional crop experiment organisations. Nevertheless, orphan crops are valued culturally, often adapted to harsh environments, nutritious, and diverse in terms of their genetic, agroclimatic, and economic niches.

Important new opportunities for improving orphan crops now exist in the tools gained through experiment on major crops and on model species, notably *Arabidopsis*. Recent achievements in the fields of genetics and genomics provide a more unified understanding of the biology of plants, which in turn can provide new opportunities for applying advanced science to orphan crops. A large discrepancy exists between the potential role of orphan crops in improving food security and the small amount of attention they have received.

Justifying significant investments orphan crop improvement requires a shift in investment from individual crops to whole sets of crops with common genetic structures. This process is identifying the correct balance of scientific investments between major and orphan crop development in poor countries so that the spillover effects from major crop are maximised and the scientific foundations for orphan crop are ensured. At the same time, attention must be paid to conventional breeding and delivery in both major and orphan crops. Crop improvement entails far more than biotechnology, and biotechnology entails far more than transgenic applications of Bt and herbicide resistance, which have been most commonly reviewed in the policy literature.

ORPHAN CROPS IN A DEVELOPMENT CONTEXT

Population growth over the next 30 years will be concentrated almost exclusively in the developing countries, where more than 1 billion people currently live on less than US$ 1 per day, more than 800 million people are undernourished, and 200 million children are underweight. This poverty is worst in rural areas where agriculture is the leading source of incomes and employment. The world's poorest regions are typically those where agricultural investments by the public and private sectors are extremely low. Unless some mechanisms can be found to stimulate agriculture, the outlook for these poor societies is bleak.

The role of agriculture in food security extends far beyond growth in crop yields and total production. Agriculture promotes food security primarily when it contributes to incomes and productive employment. Enhancing food security in the poorest regions requires investments in scientific study and training and the transfer of knowledge and technology to ensure wise management of resources and sustained capacity for growth. Moreover, food security dictates a focus on poor people's crops: subsistence and marketed crops grown in marginal areas where the poorest segments of the rural population are concentrated.

Few tabulations of the importance of orphan crops exist. Wheat, rice, maize and soybeans each occupy more than 70 million ha globally per year. Collectively, they cover 580 million ha and generate on the order of US$ 300 billion in gross value annually. Approximately two thirds of their combined area is within developing countries. Remarkably, these four crops also supply an average of about 1360 kcal of energy and 33 g of protein daily to individuals in poor counties.

Twenty-seven orphan crops within developing countries occupy areas of between 0.5 and 38 million ha. None is on the scale of wheat, rice, or maize, yet they total some 250 million ha. In addition, there are 70 million additional hectares planted to fruits and vegetables. Orphan crops take on even more significance within regions. Sorghum and millets are more important than rice and wheat, both in area and in contributions to the diet.

Similarly, roots and tubers play a dominant role, providing more than 400 kcal of energy per person per day. Given that 38% of Sub-Saharan Africa's total population is undernourished and that the number of undernourished children in that region is forecast to increase by 39% by 2020, the issue takes on added urgency. Important questions remain regarding the types of investments that are needed for individual crops or crop groupings.

Minimum investments in biotechnology approaches for individual orphan crops are likely to be at least as large as those for major crops for a particular trait. As a result, the economic incentives to invest initially in biotechnology for major crops seem clear. Once such investments have been made, however, incremental investments that extend the science to orphan crops could be extremely beneficial to some of the world's poorest populations, an opportunity often ignored.

Molecular Tools for Germplasm Improvement

Over the past few decades, large investments have been made in biotechnologies for the major crops in industrialised countries. Molecular approaches have been widely used to describe and manipulate crop genomes and to better understand the genes and biochemical pathways that govern traits such as yield, disease and pest resistance, abiotic stress tolerance, plant architecture, and quality.

Many national agricultural research programs (NARs) in developing countries have also made some investment in molecular technologies as applied to the major crops, such as rice and maize, but the extent of that investment varies sharply across countries. Nations with strong NARs are generally those with strong economies, such as China, India, and Brazil. Even in those countries, however, biotechnology investments are rarely applied to orphan crops.

More than a decade of investment has developed powerful technologies for major crops, making genetic information for crop improvement more accessible. With this technology, DNA-based approaches have been applied in two broad areas of molecular breeding. The first, marker-assisted breeding, is aimed at enhancing the power of conventional genetic analysis and manipulation. The second is transgenics.

The first step of molecular breeding uses molecular markers as tools to detect the extent and structure of genetic variation, providing insights into the diversity of crop varieties and potential contributions offered by their wild relatives. Diversity seen at this level can be used to inform the selection of parents with a high likelihood of producing novel progeny in a crossing program.

Molecular tools are also used in the analysis of inheritance of key crop traits, including those that are subject to complex inheritance due to the involvement of numerous genes. Once the genetic basis of a trait is well understood, molecular tools can be used to select for specific desired genes or combinations of genes.

The second avenue involves the transfer of genes from one genotype to another. This approach, often referred to as "genetic engineering", may utilise man-made genes, or natural alleles. Thus, truly novel traits can be added to a crop. The two broad areas of molecular breeding converge when genetic engineering is applied to enhance the efficiency with which native genes are moved within a gene pool.

Genome Analysis and Molecular Breeding

Evidence indicates that all plant genomes have a great deal in common in their gene content, biochemical pathways, and chromosome organisation. Different plant taxa have different versions of the same genes at a given position or locus in a genome, but the order of loci is conserved to varying degrees across even distantly related crops, a phenomenon known as synteny. Because plant genomes are somewhat similar, investment in "model" species, chosen because they are easily studied, has paid off. Models have been selected for each of the most important plant groups, and research has been accelerated through large-scale projects.

The flowering plants can be divided into two main groups: monocots and dicots. The former includes the grasses, which in turn include the major cereal crops. The latter include the legumes, many roots and tubers, and vegetable crops. The genome of the model dicot species *Arabidopsis thaliana* has been fully sequenced, and a great deal of basic research has been done on this species. More recently, the draft genomic sequence of the rice (*Oryza sativa*) genome was made available, which is useful both for rice and its monocot relatives.

Ninety-eight percent of the proteins found in maize, wheat and barley are also found in rice. Although the progenitors of rice and *Arabidopsis* diverged 150 to 200 million years ago, more than 80% of the genes that have been documented in *Arabidopsis* have also been found to have related genes in rice. The genes involved in many biochemical pathways and processes are very similar across the plant kingdom. Functions such as gene regulation, general metabolism, nutrient acquisition, disease resistance, general defense, flowering time and flower development are largely conserved across taxa.

Comparative mapping studies have revealed that the genomes of plant species within families are syntenous—that gene order is conserved for chromosomal segments. Given the similarities among crop genomes, particularly among plant species within a family, it seems possible that research on major crops or model species would benefit a substantial number of related crop species in the same families. The potential spillover benefits are likely to differ, however, across taxa and region depending on the particular biology of the species, the constraints faced by farmers, and seed systems.

Genetic Diversity

The use of molecular techniques for analysis of genetic diversity and the structure of germplasm, the first conceptual step in marker-assisted breeding, has been fruitful for many species. This type of research can lead to better conservation of crop genetic diversity, more efficient selection of accessions for phenotypic characterisation, and the identification of useful variants for conventional plant breeding.

Some molecular techniques (e.g., amplified fragment length polymorphism, or AFLP) can be applied to any species, without the need for crop-specific DNA sequence information. Such analysis can be conducted in any laboratory for which minimum levels of basic training, supplies and equipment are available. Even with the most seemingly straightforward techniques, however, it is difficult to achieve smooth and productive operations under the conditions encountered by many scientists in developing countries. Administrators often err in thinking that investment in facilities alone is sufficient for scientific success, and they neglect the substantial costs incurred for training, consumables, and the diversion of key personnel from field-based efforts. Even when researchers are able to produce good-quality molecular data, they may lack access to expertise in some of the analytical procedures needed to interpret the results.

This area of research provides a clear example of one in which spillovers from investments in facilities, human capital, and supplies for major crops could be quite large for orphan crops. Once researchers are well trained in the techniques and have the necessary equipment, they could apply their skills to a wider range of crops, or they could help to train others to analyse genetic diversity of orphan crops.

Marker-assisted Selection (MAS)

Marker assisted selection (MAS) is the identification of DNA sequences located near genes that can be tracked to breed for traits that are difficult to observe. MAS can be relatively straightforward for genes conditioning large phenotypic effects, such as strong effects on disease resistance, and its use for germplasm improvement is beginning to prove successful for some types of field applications for major crops. Two new rice varieties with bacterial blight resistance were introduced commercially in Indonesia in 2002.

Rice lines carrying multiple resistance genes are being produced by several national programs following technical investments at the International Rice Research Institute (IRRI). Strong national programs, such as those in China, are effectively using MAS in some cases, e.g., improvement of quality traits in rice and improvement of fiber strength in cotton.

In "single large-scale MAS", selected genes are fixed by marker-assisted selection, and the constitution of the rest of the genome is determined by conventional field-based breeding. This approach has been used to develop virus resistance in maize, rice, and beans. In recent cost analyses of marker-assisted versus conventional selection, the cost of MAS for a recessive mutant gene associated with quality protein maize was found to compare favorably with the conventional screen.

As basic experiments on key plant traits advances and as marker technologies continue to become increasingly simple and powerful, more applications of MAS are anticipated with the potential benefit of faster varietal release. Morris et al. showed positive cost-benefit results of MAS, and there are many similar unpublished examples from the private sector. In the published literature the cost-benefit results of MAS are mixed, and many publications only report work aimed at eventual application of

MAS as opposed to field-level use of MAS for practical crop improvement.

Due to the relatively high cost of MAS compared to conventional field screening, MAS is likely to be practical for orphan crops in the poorest countries only if partnerships with well-funded research groups allow free use of existing technologies. It is important, therefore, to assess various traits and trait categories for which there has been some proof-of-concept in non-orphans for MAS before considering what would be required to move the benefits to a broader set of plant species.

In viewing cases in which genetic research has allowed for progress in MAS, it is important to distinguish between two types of traits: those conditioned by single genes with large effects showing simple inheritance (qualitative traits), and those showing complex inheritance (quantitative traits). Important plant traits that frequently show simple inheritance include some resistances to diseases and insects and some important characters that relate to plant growth and development. It is relatively easy to locate these genes with precision, and in many cases such genes have been cloned. It is often straightforward to select for these genes in a conventional breeding program, in which case MAS does not necessarily offer substantial benefits.

There are cases, however, in which markers are particularly beneficial; for example, when a disease resistance gene must be identified but the pathogen is not present at the breeding site, or when a gene is to be selected at the seedling stage for a trait that is only expressed later in development.

Most important traits are governed by multiple genes, each having relatively small effects. These "quantitative traits" have been difficult to understand and to manipulate in conventional crop breeding programs. The term QTL, quantitative trait locus or loci, refers to the chromosomal regions of genes that control quantitative traits. Molecular genetic approaches began to illuminate the genetic architecture of quantitative traits. By now, the chromosomal segments associated with many traits have been identified in a large number of experiments. As estimated by Goff et al., about two thousand cereal QTLs have been mapped.

While so-called anonymous QTL-associated markers are sometimes used for marker-assisted selection in crop improvement, the utility of such markers can be limited by a high degree of imprecision in mapping desirable loci of small effects. The identification of "candidate genes" makes it possible to localise desirable variants much more precisely.

Candidate genes are genes known or suspected to be involved in conditioning the phenotype of interest, such as disease resistance. The inference that a particular gene contributes to a given trait may be strengthened if it maps to a chromosomal region that has been associated with the trait in QTL mapping experiments.

Credible candidate genes co-localising with QTLs have been identified for several traits, including quantitative disease resistance in rice, wheat, bean, and potato. A number of experiments approaches have converged to allow some genes underlying QTLs to be cloned, and to set the stage for future QTL cloning in others. This process permits both the identification of potentially useful variants of agronomically important genes and the precise selection of those alleles found to be most useful.

One of the approaches that puts QTL cloning within reach is the increasing availability of express sequence tags (ESTs), short segments of sequenced gene transcripts that provide information on genetic expression, function, and heritability. Tens or hundreds of thousands of ESTs are available for major crop species, but the number of sequences for orphan crops is quite meager. The combination of QTL mapping data and mapped EST data as a powerful tool for determining the genes that underlie quantitative traits.

Sequence data on expressed genes and on plant and crop genomes are rapidly accumulating and present powerful tools for plant science, provided that scientists are able to access and exploit the data efficiently. The sequence data sets are, in themselves, imposing and potentially cumbersome. But with the help of increasingly powerful and friendly databases, biologists and breeders can now gain access to genetic information that allows them to identify and exploit natural variation for a wide variety of crops in ways that were previously not possible.

Regulatory Genes

The genetic complexity underlying quantitative traits often makes it difficult to seek trait-marker correlations that can be used for indirect selection. For instance, a great number of genes (hundreds or thousands) are involved in plant defense against insects and pathogens. This complexity may seem defeating, but it may be possible to improve complex traits by manipulating the genes that sit at the top of the regulatory hierarchy and control large numbers of other genes. In fact, there is growing evidence that many QTLs are explained by transcription factors-genes that control the expression of other genes. The identification and utilisation of the "best" alleles of these genes is a promising avenue of research, both via MAS and via transgenic approaches. Transformation experiments enhancing the stress-responsive expression of a transcription factor that controls drought-related genes led to increased tolerance to drought, salt stress, and freezing in *Arabidopsis*.

Similarly, overexpression of a regulatory factor involved in plant defense in transgenic *Arabidopsis* led to enhanced resistance to diverse pathogens. Some of the genes involved in regulation of the defense response are also implicated in responses to other stresses. The identification of these transcription factors seems likely to play a useful role in genetic improvement for a wide range of food crops.

The use of direct gene transfer, the second tool for "molecular breeding" of crop traits, employs recombinant DNA technologies to insert one or more genes into the crop plant's genome. While most transgenic research and applications have focused on a few major crops to date, the potential to extend applications of this technology to orphan crop improvement may be significant. Promising transgenic lines for about 20 different crops, including pepper, squash, and sweet potato, are currently being field tested by public research institutions in at least 10 developing countries. These lines, which have been engineered for particular traits like virus resistance and pest control, could provide significant benefits to poor farmers who cannot afford chemical controls for crop loss.

A wide range of approaches employing transgenes is available for controlling herbivory, but by far the most extensively used method to date has been "Bt" technology. Different Bt toxins are lethal to the immature or larval, herbivorous stages of lepidoptera (adults are the butterflies and moths), diptera (the flies, including mosquitoes), and coleoptera (the beetles, including weevils), many of which are major crop pests.

Beyond the use of Bt, several additional insecticidal products are produced by plants, animals, and microbes that are also being evaluated for the control of herbivory. For example, many plants store large amounts of specialised proteins in their seeds (called storage proteins), and some of these proteins have insecticidal properties.

Insects that consume plant seeds are particularly dependent on their ability to digest seed proteins using one or more proteases, and these inhibitors have been engineered into other plants to interfere with the feeding of insects on leaves and flowers. Protease inhibitors have also found to be effective against some of the plant pathogenic nematodes.

A number of enzymes have also been tested for their possible use as insect resistance determinants. One of these enzyme groups is the chitinases, which target an essential structure of macromolecules (chitin) in insects and fungi. Three additional groups of enzymes (conifer monoterpene synthases, bacterial cytokinin biosynthesis enzymes, and cholesterol oxidase) have been designed to modify the chemistry of the plant in ways that make it less palatable as an insect food source.

Finally, in many crops, especially the grains, a major product stored in seeds is starch. Insect pests of these crops depend on their ability to digest starch using the enzymes α and β-amylases. Some plants encode a gene for an inhibitor of α- and β-amylase; these genes have been used in a variety of cases to produce insect resistant plants.

Transgenic approaches for reducing crop losses from disease have also been widely explored, with significant success in the area of plant viruses. Reports of success on transgenic technologies for controlling plant viruses abound in the literature and encompass numerous crops including rice, wheat, maize, barley, perennial rye, legumes, potato, sweet potato, several temperate vegetable species, and papaya.

Plant viruses represent a major biotic constraint on agricultural production; viral symptoms commonly include stunting, reduced yields, or death. Viruses are usually transmitted to plants via insect, beetle, nematode, or other types of vectors, and as a result, resistance is often the only effective form of viral control in agricultural systems. While genetic resistance exists in many crops, it is not always well understood or identified. For orphan crops in particular, even the most basic knowledge of resources for genetic resistance is typically lacking.

The mid-1980s confirmed that when portions of plant viral genomes, such as the viral coat protein gene, were transferred and expressed in a plant genome, resistance to the virus was observed. Many types of sequences from both RNA and DNA plant viruses have been expressed in a diverse set of crops resulting in resistance to a wide array of pathogens. More recently, scientists have turned their attention to resistance resulting from expression of genes from other sources, including mammalian antibodies, anti-viral compounds, and sequences that act to silence viral genes or silence host genes necessary for viral infection. To date, however, resistant varieties resulting from "second-generation" strategies (i.e., those involving gene silencing or expression of non-viral sequences) are not yet in use.

Nonetheless, future strategies to limit crop losses to viral disease will most likely encompass upstream research on sequences that silence plant genes and on direct gene transfer to move plant genes beyond the boundaries defined by sexual hybridization.

Genetic engineering for insect or virus resistances is a so-called "implementation technology", meaning that once the genes are available, they must be combined using an "enabling technology" required to move the genes into target crops. Thus, in principle the sequences used that encode the "pesticidal" activity, combined with appropriate regulatory sequences that work in the target crop, can be used in any crop plant. In this way, genes isolated originally for use in a major crop such as wheat or soybeans can be readily modified for use in orphan crops. The technical challenges for scientists applying these technologies to orphan crops are threefold.

The first challenge relates to the extent of knowledge about the sensitivity of the pathogens or pests of orphan crops to forms of control that have been useful in developed countries with temperate environments. Work in India on insect resistance for chickpea showed that the proteins encoded by the previously isolated proteinase inhibitor genes were at best marginally effective against insect pests of chickpea. The researchers then turned to isolation of orthologous genes from crop plants such as Asian winged bean to find more effective proteinase inhibitors.

The second challenge concerns the availability of techniques for transferring genes into orphan crops. In some cases, the transferability of gene transfer methods has been fairly easy and direct from model species to orphan crops. But in other cases, significant new and expensive work has been needed to modify-or to develop entirely new-transformation technologies for these crops.

And finally, a major limitation of the use of these promising technologies in the orphan crops to date is the paucity of approaches that can be successfully applied to the control of diseases caused by bacteria, fungi, the oomycetes, and nematodes.

GENOMICS-GUIDED TRANSGENES (GGT)

Genomics approaches and transgenic technology converge when natural allelic variation is utilised through direct gene transfer. Useful alleles can be transferred from a wild relative of a crop by conventional breeding, but it may be difficult or even impossible to recover the desired plant type in a reasonable period of time. Using "genomics-guided transgenes" (GGT), native genes or homologous genes from closely related species that modify a plant's metabolism in a manner similar to natural or induced mutations are used for direct gene transfer. Dominant alleles that confer important agronomic traits of interest (e.g., pest and disease control, drought tolerance) but that are scarce in breeding populations can be identified and inserted into breeding lines, based on knowledge gained from model gene studies. An example of this approach is the use of the *Xa21* gene, which provides resistance to the bacterial blight disease of rice. *Xa21* was originally transferred from a wild relative to cultivated rice by conventional breeding.

The gene has been moved among genotypes by MAS and has also been cloned and utilised through genetic engineering. GGT can comprise a wide diversity of genes, although many of the modified traits using this approach to date were already familiar, such as male sterility, seedless fruits, delayed spoilage, and dwarf stature.

Given the range of molecular tools, what are the promising applications for orphan crops? The answers can best be explored when related major crops have received substantial research investment. The most obvious example is the grass family, or Gramineae, the most economically important plant family. There are around 10,000 graminaceous species, which include the world's most important cereal crops. Knowledge of the taxonomic structure facilitates the use of information from the more studied species to the orphans.

There are five sub-families in the grass family. Rice is a member of the ehrhartoid sub-family and has the smallest genome. The pooids include wheat, barley and rye. The panicoids include maize, sorghum, and pearl millet, all of which are important staples in sub-Saharan Africa and other areas. The chloridoids include the Ethiopian staple tef, as well as finger millet. The bambusoids include bamboo.

Within the grass family, genomes vary substantially. While the genomes show similarities in gene organisation, there is evidence of significant variation in genes, chromosomes and genomes. Based on a framework of restriction fragment polymorphism (RFLP) maps, the cereal chromosomes show gross synteny. At a finer level, however, there are many deviations from synteny, such that the predictive power between taxa ranges generally between 50 and 75%. Even between maize lines, the gene content of a given chromosomal region vary.

Some of the genes affecting important traits such as seed size, seed dispersal, and flowering time have been shown to exist in corresponding sites in the genomes of rice, sorghum and maize. Several QTLs conditioning resistance to the blast disease pathogen were found to exist in the same locations in both rice and barley.

In many cases, therefore, knowing the location of a gene or group of genes affecting a key trait in one species will be useful for efficient identification of genes that are of interest for another trait. If the actual genes are known, then the gene discovery process is less dependent on synteny, as the genes can be traced by sequence similarity. Similar examples are found for a range of plant families. For instance, the Solanaceae include many vegetable species that are extremely important as food and cash crops. For tomato, a substantial amount of genetic and genomic information is now available. The situation for potato and pepper is intermediate.

The chromosomes of these species are syntenous, and the genes for various characters have been found to reside at syntenous locations. For various African indigenous vegetables (AIVs) in the Solanaceae, there is virtually no information available, but information from the better-studied solanaceous crops would doubtless be an asset if an investment were made in the AIVs. A similar situation exists for the legume family. This mapping, while only partial, helps to identify niches of opportunity for biotechnology spillovers into orphan crops. Much greater challenges exist for orphan crops like quinoa and some roots and tubers that have no close model or major crop relatives.

Despite identifiable niches, the task of implementing biotechnology approaches to germplasm improvement is not simple, particularly in poor regions. The process of analysing the diversity of genes related to a trait of interest, and identifying and marking the useful alleles for selection, requires good access to current scientific literature and the internet, with at least modest bioinformatics capacity.

The best promise for implementing a specific biotechnology strategy for germplasm development in orphan crops holds when research on a particular trait is pursued across a group of related species, one or more of which is not an orphan. The most rapid progress is likely to be made when there is a model crop in the same family as the orphan crop.

In the sweet potato case, weevils are known to reduce yields by 60%, and in some cases by 100%. Average yields, currently at 4.4 tons/ha, would thus rise to over 7 tons/ha with fully effective resistance. Viewing potential success more conservatively, a 25% increase in yields from weevil resistance would raise the average to 5.5 tons/ha.

Given a sweet potato area of 570,000 ha in Uganda and a price of US$ 88/ton, the gain from achieving Bt-derived weevil resistance on half the area would be over US$ 27 million per year. If the technology could be distributed throughout the Sub-Saharan Africa on one-half of the sweet potato areas, the gross annual benefits would be about US$ 121 million. Twice the initial investment would be needed to move the resistant cultivars into the field in Uganda (US$ 10 million), and that five times the initial amount would be needed to disseminate the technology throughout the Sub-Saharan region (US$ 25 million). Even with these expenditures, the net payoffs would still be large. Moreover, because sweet potato is a vegetatively propagated crop, the biosafety risks of introducing Bt in this case would be much lower than for maize or cotton.

Investing in marker assisted selection for blast resistance in finger millet could also prove successful within the next decade, particularly given the gains made in identifying genes and QTLs for blast resistance in rice. Finger millet yields are typically reduced by 35% or more as result of neck and finger blast. Successful breeding for blast resistance could increase average yields from 1.3 ton/ha currently to over 1.75 ton/ha.

A more conservative estimate of potential success would be to achieve a 15% gain in yields from resistance (to 1.5 tons/ha on average) on one-half of the total area. Given an area of 3 million ha planted to finger millet in India, another 1 million ha in Africa, and a price of US$ 96/ton, the gross annual benefits would be more than US$ 38 million globally. Again, even if large costs were incurred to move the technology into the field, the net pay-offs would still be large.

Greater challenges exist with a crop like tef, where the technology path leading to the creation of dwarf varieties is more obscure. In all cases, however, there are benefits associated with the investments, including the enhancement of scientific capabilities and human capital development. The probability of success, the potential financial returns, the collateral benefits, and the constraints on technology adoption should all be weighed.

Success in the laboratory or experimental plot still needs to be integrated into a much broader scientific process in order to be successful in rural communities. A technological continuum exists for germplasm improvement, which ranges from simple selection techniques, to conventional breeding approaches with farmer participation, to biotechnology developments in advanced laboratories. Unfortunately, the continuing decline in funds for practical crop improvement, particularly in the international public sector, results in a fair amount of investment in research, and a weak application to practical problems. Applying molecular techniques to basic research in plant biology has led to a much deeper understanding of the genetics of key traits, but not always to a clear idea of how plants can be improved.

Appropriate use of biotechnology depends on the agricultural problem at hand, the biological properties of the crop, and the economic and social infrastructure that supports crop research. The best chances for harnessing the gains from biotechnology exist when the science is integrated into breeding efforts, farm management, and seed production and distribution. Even with integration, the benefits from advanced science depend critically on the institutional, human capital, economic, and political context of the recipient countries.

Useful generalisations about potential field-level applications of biotechnology for orphan crops are made difficult by the great diversity of crops and socioeconomic settings. Nevertheless, orphan crop improvement is likely to benefit from attention in three areas: dispelling the myth that biotechnology is equivalent only to genetic engineering applications of Bt insect resistance and herbicide resistance; creating new incentives for public-private partnerships; and fostering new institutional arrangements that combine efforts across whole sets of major and orphan crops.

Many people in the policy community—and in the general public—believe that agricultural biotechnology encompasses only transgenic approaches to insect and weed resistance. This focus on GMOs has pitted consumers against industry, Europe against the US, and environmentalists against each other as they debate the relative costs of pesticide use and genetic engineering. While the debates are valid in their own right, they overlook the fact that many new genetics and genomics techniques do not involve the insertion of foreign genes into plants.

The use of molecular tools to understand the genetic basis of crop traits for indirect selection and breeding is beginning to offer an alternative to traditional transgenics. These technologies will help alleviate some of the constraints surrounding intellectual property rights and biosafety that limit the dissemination of GMOs in developing countries. Many developing countries have been reticent to employ GMOs, lest their use interfere with trade and aid. It is the responsibility of agricultural policy analysts and practitioners to help redirect policy and public discussions on the future of biotechnology. The point should focus foremost on the needs of poor farmers and consumers, and the tradeoffs they face. To the extent that conventional breeding can be facilitated by advanced genetics techniques, biotechnology should be encouraged as one tool in the agricultural development package.

Public-Private Partnerships

If newer forms of biotechnology are to be applied, and especially those involving spillovers from one crop to another, partnerships will be necessary. Many partnerships exist and can be built upon, especially among universities and national agricultural research institutions (NARs). Two types of linkages that transcend the traditional university-NAR relationships are also likely to be key: honest broker organisations that bring together needs and skill sets, and the private sector that is the home for much of the relevant technology. The private sector is also essential in moving crops from the lab to the field. In many cases, the public sector has a comparative advantage in discovery, while the private sector has a comparative advantage in overcoming regulatory barriers and distributing seeds.

Creating new public-private partnerships will be essential for managing intellectual property issues. Ten companies, which own more than three-fourths of the agricultural patents in the

US, are likely to be central in many of the biotechnology applications. Significantly, Dow, DuPont, Monsanto and Syngenta have already agreed to provide seed varieties, patent rights, and laboratory knowledge to African countries through the African Agricultural Technology Foundation (AATF).

The resulting spillovers into orphan crops could be especially important, although for this to happen, AATF will probably first need to succeed with a major crop such as maize. The private sector controls patents on transformation technologies for genetic manipulation as well as on the products themselves. In order for some of the newer forms of transgenics applications to work (i.e., those that do not involve Bt), patent rights on these transformation technologies will have to be relinquished not only for research, but also commercialisation.

The Center for the Application of Molecular Biology to International Agriculture (CAMBIA) has a major focus on securing patent rights for the public sector, particularly in developing countries. In addition, a consortium of US universities convened by The Rockefeller and McKnight Foundations has recently launched an initiative called the Public-Sector Intellectual Property Resource for Agriculture (PIPRA), designed to make agricultural technologies accessible for humanitarian uses in developing countries.

Intellectual property is not the only factor discouraging private and public partnerships. For much of Africa and parts of Asia and Latin America, the lack of biosafety protocols and their enforcement creates serious problems of legal liability. Without such protocols, external public and private organisations will be exceedingly reluctant to become involved with orphan (or major) crops, and for very good reasons. No responsible organisation wants to be a party to, for example, an "escaped" pathogen or an inserted gene that out-crosses into a weedy relative. Remy et al. say that transformed bananas resistant to two devastating diseases in Rwanda and Burundi have been developed in Belgium. Because of the lack of biosafety protocols, the transformations remain in the laboratory.

Issues of intellectual property rights and biosafety are not confined to the transfer of science to orphan crops; major crop improvement programs face similar problems in developing countries. From a policy perspective, new programs will need to be developed in the legal and safety areas-in many cases with the help of external aid and consultation-to facilitate the adoption of technology and its use in farmers' fields. In terms of policy priorities, the establishment of legal and safety measures is often more important that allocating additional funds to biotechnology for crop research, lest new technologies simply remain on the shelf. These constraints will likely be much lower when molecular markers and mapping techniques aid conventional crop breeding. Nonetheless, private sector cooperation may still be helpful in distributing seeds.

For major crops, such as wheat and rice, much of the linking of research with farmers has been performed by Centers of the CGIAR. Limited inroads on millets, cassava, bananas, and a few other crops have been made by additional CGIAR Centers, but there are large numbers of orphan crops where virtually nothing has been done. It is on this point that the new AATF instituted by the Rockefeller Foundation holds considerable promise. Its mission is to help design the relevant templates, protocols, and procedures that will lower the transaction costs of applying biotechnology to major and orphan crops in Africa. By doing so, it will provide a model for other regions.

The McKnight Foundation's Collaborative Crop Research Program is also aimed at enhancing the transfer of science from major to orphan crops and training scientists from poor countries in advanced genetics and genomics methods. Finally, the biotechnology programs being developed and promoted in Dutch, Swiss, and US aid agencies are contributing to progress in orphan crops. Special programs and new incentives within the scientific and development communities should be further encouraged to achieve widespread spillover benefits for poor farmers.

8

IMPROVING STRESS TOLERANCE IN PLANTS

Crop yields are frequently lowered by biotic and abiotic stresses, and one of the most effective strategies to improve agricultural output is to breed or engineer plants tolerant or resistant to stress. Initial research on molecular manipulation focused on production of enzymes that detoxify reactive oxygen species, such as superoxide dismutase. Reactive oxygen species are induced by most types of stresses. Since then, many approaches have been pursued to produce plants tolerant to a broad range of stresses.

When a plant is exposed to stress, survival mechanisms turn on to reduce damage. The balance between stress and survival signals determines the level of damage suffered by the plant. Among key molecules studied in relation to stress are reactive oxygen species (ROS), which, due to their high reactivity and therefore toxicity, have been called the unwelcome companions of aerobic life. Under stress conditions, the steady state level of ROS usually increases, and it has been hypothesized that ROS (specifically hydrogen peroxide) might also act as messengers, turning on stress-related genes. Recent progress in plant signal transduction reveals that ROS are more than just the misfortunes of an aerobic life. They also play a regulatory role in the plant's physiology.

Since it is not very practical to completely avoid growing crops in conditions that are stressful, scientists focus on engineering plants to be indifferent to stress. Marc DeBlock and colleagues say a strategy of improving stress tolerance in plants by maintaining the plant's energy homeostasis under stress. When plants are exposed to a stress signal, they expend a lot of energy and exhibit enhanced respiration rates. This is partially due to a breakdown in the NAD^+ pool caused by the enhanced activity of poly(ADP-ribose) polymerase (PARP), which uses NAD^+ as a substrate to synthesise polymers of ADP-ribose.

This poly(ADP ribosyl)ation is a posttranslational modification of nuclear proteins that seems to be initiated by oxidative and other types of DNA damage. Stress-induced depletion of NAD^+ results in a similar depletion of energy, since ATP molecules are required to resynthesise the depleted NAD^+. DeBlock and coworkers show that plants with lowered poly(ADP ribosyl)ation activity appear tolerant to multiple stresses.

First, the researchers demonstrated that inhibiting PARP activity via chemical inhibitors (3-methoxybenzamide, nicotinamide, and isonicotinamide) protects plants from oxidative stress. To produce transgenic stress tolerant plants, they transformed Arabidopsis and oilseed plants with double-stranded RNA constructs of the *parp1* and *parp2* genes. Use of double-stranded RNA is now a widely accepted strategy for silencing endogenous genes. In this way, the group down regulated *PARP1* and *PARP2*.

Arabidopsis and oilseed plants were transformed with dsRNA constructs containing the 5' end of Arabidopsis *AtParp1* (hairpin*AtParp1* or hp*AtParp1*) or *AtParp2* (hp*AtParp2*) in the stem structure. The AtParp lines showed enhanced tolerance to stress, such as heat and drought.

Managing the Economics of Energy Under Stress

Most stresses interfere with mitochondrial function. Stress deregulates the physiology of the plant and causes NAD^+ breakdown, ATP overconsumption, and enhanced respiration. These reactions deplete the energy of the plant, cause the production of reactive oxygen species, and consequently induce cell death.

To demonstrate the high energy efficiency of hpParp lines, the researchers cultured hypocotyl explants of control and hpParp plants on medium containing 0.06 M glucose as the sole carbon source. While the control plants were necrotic and showed poor callus formation, hpParp lines remained green and showed vigorous callus. Germination on medium containing 2-6% glucose resulted in no difference in seedling growth between the hpAtParp lines and the non-transgenic control line. This indicates that the differential callus formation and survival of the explants on glucose medium between the stress-tolerant hpAtParp lines and the control line is not due to an altered sugar sensing, but rather due to efficient energy metabolism in the hpParp lines.

Poly(ADP-ribosyl)ation plays an important role in the recovery of proliferating cells from certain types of DNA damage, and this has been linked mechanistically with an involvement in DNA base-excision repair. It is therefore not clear whether plant parp genes have a similar function as animal parps, in terms of maintenance of genomic stability. If they do, downregulating them may have negative effects to the growth of the plant.

Breeding Greater Drought Tolerance into Wheat

The emergence of the agricultural biotechnology industry at the end of the 20th century brought with it products that fall into two basic categories: herbicide resistant and insect resistant crops (i.e., maize, soybean, and cotton). In the late 1980s and the 1990s, attempts were made to develop a range of other products whose qualities were to include improved fungal disease resistance, starch composition, fruit quality, and so on. As the agricultural biotech industry matured, standard product development processes were adopted. For agricultural products, the commercialisation of initial discoveries generally takes 8-12 years, from gene discovery to target validation. As genomics applications emerged, the agricultural biotechnology industry recognised the potential to identify candidate intervention points for commercial traits.

In reaction to stresses, plants adjust themselves at the levels of morphology, phenology, physiology, and biochemistry. Because these responses are presumed to be regulated by genes, efforts in recent years have focused on the isolation and characterisation of genes induced by stresses. Among stress-induced genes isolated to date, several major groups have been targeted for improving abiotic stress resistance in plants. These include genes encoding enzymes for the biosynthesis of compatible compounds, enzymes for scavenging active oxygen species, heat shock proteins (HSPs), late embryogenesis abundant (LEA) proteins, enzymes modifying membrane lipid saturation, transcription factors, and proteins required for ion homeostasis.

Among abiotic stresses, drought is the most important from an economic standpoint and likely the most intractable to breeders' efforts. Diffficulties in breeding for drought tolerance are compounded by an incomplete knowledge of the genetic and physiological bases of yield in water-limited conditions. To overcome the minimal response to direct selection for yield under drought conditions, substantial efforts have been directed toward manipulation of morpho-physiological traits that influence drought adaptation through escape, avoidance, and/or tolerance mechanisms. However, this indirect selection strategy has been successful in only a limited number of cases.

More positively, in recent years several genes that are responsible for different traits associated with drought tolerance have been isolated and characterised. In another branch of the biotech discipline, an increasing number of experiments have strived to map quantitative trait loci (QTL) affecting drought-related traits and yield in major crops exposed to water deflcit. In a limited number of cases, markerassisted selection (MAS) has been used as an integral component of breeding strategies to incorporate target traits and to increase tolerance to drought.

More recently, bioinformatics and the flood of information generated by genomics platforms have added new dimensions for understanding the role and function of genes governing the response to drought. Despite these impressive technological breakthroughs, the overall impact of MAS and genomics on the release of drought resilient cultivars has thus far been negligible.

As water shortages approach critical levels around the world, particularly in developing countries where more than 60 percent of the inhabitants already live under precarious conditions with no access to safe sources of fresh water, we clearly need to provide quick solutions to the multi-faceted dilemma of water shortages. One promising approach is to integrate useful drought tolerance traits through genetic engineering into lines or varieties at the advanced breeding level for drought tolerance.

Genes carrying these traits are now coming on line and, applied together with conservation agriculture techniques, may offer developing world farmers a response to water-limited conditions that is sustainable, economical, and readily adoptable.

The most widely used promoters in development of transgenic plants are constitutively expressed, i.e., they are turned on all the time and throughout the plant. In cases where gene expression needs to be tailored to a specific organ or a specific time, such a promoter is not a good choice, especially for the stress-induced genes. This is because the constitutive expression of a stress-induced gene may have serious penalties with respect to energy loss or other adverse side effects. Thus, a stress inducible promoter should be considered when the transgenic plants are targeted to deal with abiotic stresses. Points of the cis-elements responsible for stress induction is necessary to temporally and spatially target the transgenes. Several studies provide successful examples of the use of an inducible promoter.

Ability to Test Transgenic Plants under Field Conditions

Overexpression of stress-related genes has afforded some stress protection in transgenic plants. However, the results are not always consistent and can even be conflicting. Due to the multigenic nature of stress tolerance, comprehensive physiological and biochemical testing of transgenic plants under stress conditions must be conducted, which requires a careful evaluation of the methods for assessing stress tolerance, especially osmotic stress. Desiccation and salt stresses applied by most researchers are 'shock' treatments. For most crops, drought tends to develop slowly as the soil dries.

Plants that are subjected to drought conditions in this gradual manner accumulate solutes that maintain cell hydration and undergo complex adjustments in their morphology and photosynthetic characteristics. Thus, to ensure that the responses of the transformed plants to water stress treatments are agronomically relevant, plants must be subjected to the same drought regime that crops experience in the field.

Ability to Transfer Multiple Genes

While introduction of the key abiotic stress resistance/tolerance genes into plants increased their stress resistance in some recent experiments, simultaneous transfer of several genes will be the likely next step to achieve practical levels of plant resistance/tolerance. As abiotic stress resistance is polygenic, plant engineering will require manipulation of complex metabolic or regulatory pathways involving multiple genes. If osmoprotectant-producing, transcription factor expressing, ion homeostasis-maintaining, in the case of salt, and antioxidant enzymatic activities are all incorporated into a single cultivar, there is a strong possibility that they could work in concert to overcome concurrent abiotic stresses. This could be achieved either by transformation with multiple genes or by crossing plants containing different stress tolerance genes.

Developing Sustainable Agriculture Systems

Agronomic measures to reduce yield losses related to drought stress are a key aspect of an integrated set of technologies called conservation agriculture (CA). The overall objective of CA is to increase yields and sustain crop production and environmental quality through more efficient use of natural resources and chemical inputs.

International and national research has contributed to the development and testing of CA technologies suitable for local agroecological and socioeconomic conditions in Africa, Asia, and Latin America, and a lot of expertise is available. To optimally exploit the combined benefits from agronomic and Genetic Engineering (GE) technologies to reduce yield losses due to drought stress, collaborative efforts between genetic engineers, breeders, physiologists, and agronomists are required. Moreover, integration between agronomic and GE approaches is pivotal to ensure the sustainability of the technologies with respect to the conservation of soil fertility and natural resources in the long term. Three basic principles of CA are:

(i) zero or reduced tillage;

(ii) retention of soil cover with crop residues and/or cover crops that form a barrier to water loss by evaporation and runoff; and

(iii) crop rotations including, where possible, biological nitrogen-fixing crops.

Such systems are increasingly adopted to stop or reverse soil degradation resulting from unsustainable management practices. Today CA is used on more than one-third of the cropped areas in Brazil and Argentina, and about 70 million hectares worldwide. CA systems demonstrate clear benefits in terms of reduced labor and/or fuel inputs.

Other advantages include reduced soil erosion, improved water infiltration and soil structure, increased soil fertility, soil organic matter accumulation, carbon sequestration, soil water-holding capacity, water use efficiency, soil biodiversity, and resilience to climate change.

The combined effect of all these factors is claimed to increase and sustain agricultural productivity with lower water, energy, and labor inputs. Increased adoption of CA technologies, therefore, not only enhances food security for millions of smallholders in the developing world, but also reduces detrimental environmental effects.

CA management practices for rain-fed conditions have been developed and their effects investigated in some welldesigned studies. Results from the long-term management trials of CIMMYT in the subtropical highlands of Central Mexico indicate that small-scale maize and wheat farmers in this region may expect yield improvements of 67 and 84 percent for wheat and maize, respectively, after adoption of zero tillage, appropriate rotations, and retention of sufficient residues as compared to current practices of heavy tillage, monocropping, and burning or removal of residue.

An additional CA technology generating considerable interest is raised permanent beds, where the benefits of zero-tillage are combined with a bed and furrow system. The practice of planting crops on beds/ridges that are formed between furrows used to supply irrigation water is widely applied in many (semi)arid regions under irrigated agriculture, although generally beds are tilled and reshaped before each crop cycle. However, permanent bed systems have also been shown to be a promising technology in water-limited, rain-fed environments, because they tend to conserve rainfall, prevent runoff, and provide time for water to infiltrate, with the surplus advantage that more varied weeding and fertilizer application practices are possible.

Use of transgenics is a large challenge for public breeding programmes, but also one of the biggest opportunities. A more efficient and sustainable agriculture depends on plant varieties and cropping systems with increased resistance to diseases, pests, and other environmental stresses. Transgenics are a robust source for these traits, although target traits should be identified in close collaboration with agronomists.

CA technologies aimed to establish sustainable, water-conserving cropping systems may require specific tolerances/ resistances or traits, for example herbicide resistance or tolerance against plant pathogens and drought that are favoured in zero-till systems.

Abiotic Stress Tolerance in Grasses

Abiotic stresses, notably extremes in temperature, photon irradiance, and supplies of water and inorganic solutes, frequently limit growth and productivity of major crop species such as wheat. Inaddition, more than one abiotic stress can occur at one time. High temperature and high photon irradiance often accompany low water supply, which can in turn be exacerbated by subsoil mineral toxicities that constrain root growth. Furthermore, one abiotic stress can decrease a plant's ability to resist a second stress. Low water supply can make a plant more susceptible to damage from high irradiance due to the plant's reduced ability to reoxidize NADPH and thus maintain an ability to dissipate energy delivered to the photosynthetic light-harvesting reaction centers.

If a single abiotic stress is to be identified as the most common in limiting the growth of crops worldwide, it most probably is low water supply. However, other abiotic stresses, notably salinity and acidity, are becoming increasingly significant in limiting growth of both forage grasses and the cereals.

Globally, low temperature also is a major limitation of plant growth, and this has a major impact on grasses via, for example, vernalisation and low temperature damage at anthesis. In this focus issue, there are articles addressing three aspects of these abiotic stresses.

Traditional approaches to breeding crop plants with improved abiotic stress tolerances have so far met limited success. This is due to a number of contributing factors, including: (1) the focus has been on yield rather than on specific traits; (2) the difficulties in breeding for tolerance traits, which include complexities introduced by genotype by environment, or G × E, interactions and the relatively infrequent use of simple physiological traits as measures of tolerance, have been potentially less subject to G × E interferences; and (3) desired traits can only be introduced from closely related species.

Most cereals are moderately sensitive to a wide range of abiotic stresses, and variability in the gene pool generally appears to be relatively small and may provide few opportunities for major step changes in tolerance. Of potentially larger impact on abiotic stress tolerance is the use of genetic manipulation technologies to generate such step changes. Having said this, more immediately achievable, if modest, increases in tolerance may be introgressed into commercial lines from tolerant landraces using marker-assisted breeding approaches, facilitated by recent breakthroughs with positional cloning that are likely to enable identification of extant tolerance genes within cereal germplasms.

Of course, the sequencing of the rice (*Oryza sativa*) genome provides an invaluable resource for work on rice and, by exploiting syntenic alignment with many other grasses, facilitates fine mapping in the unsequenced genomes of many other grasses.

It is exploitation of this latest resource that, combined with steadily increasing transformation frequencies for many grasses, is making the functional genomics approach to the manipulation of abiotic stresses in grasses increasingly tractable. The need to use a model plant such as Arabidopsis (*Arabidopsis thaliana*) for such work is steadily decreasing, and will continue to do so, as the principles uncovered in this model organism are refined by knowledge gained in the plants that are the ones in which this knowledge needs to be applied.

Furthermore, in addition to the obvious fundamental differences in development and anatomy between monocotyledons and dicotyledons, many of the mechanisms of tolerance to abiotic stresses can have fundamentally different characteristics between these two major plant groups, so transferring knowledge from Arabidopsis to the major crops often is not possible. When grown in saline soils, many dicotyledonous halophytes accumulate much higher concentrations of Na^+ in their shoots than monocotyledonous halophytes, a feature that may be related to the observation that succulence is observed more commonly in dicotyledons than monocotyledons, particularly the grasses.

The possibilities for increasing tolerance to abiotic stresses are enormous, although it is notable that the actual production of transgenic plants with demonstrably improved abiotic stress tolerance has been slow. Knowledge from work in grasses can be applied for increasing abiotic stress tolerance in commercial lines. The use of $\delta^{13}C$ assays for screening wheat lines to generate the new drought tolerant line Drysdale is an excellent, if rare, example of the use of new technologies developed from pure plant physiology in otherwise traditional breeding programmes to increase abiotic stress tolerance.

Step changes in tolerance may arise from the introduction of *de novo* characteristics that are apparently completely absent from a particular gene pool. No wheat variety apparently has the ability to synthesise the sugar alcohol mannitol. This can be accumulated to large concentrations in some plants, where it can act as a compatible solute, decreasing the osmotic potential in the cytoplasm and thus increasing the ability of the cytoplasm to retain water in the face of reduced water supply. However, when the bacterial gene encoding the key enzyme to allow mannitol synthesis was inserted into wheat plants, with expression driven constitutively, it usually caused severe stunting of plants and sterility.

Interestingly, in the lines that synthesised very low concentrations of mannitol, vegetative growth was inhibited less by either the addition of 150 mM NaCl to the liquid medium in which they were grown or by reduced watering of soil-grown plants. There appeared to be no reduction in growth of these plants in unstressed conditions. It will be interesting to know the effects of this manipulation on seed yield.

Manipulation of processes involved in some of the later responses to stress, related more to processes of damage limitation, rather than damage prevention. A more preemptive defense against abiotic stress could involve processes involved in the early detection of and response to stress. This approach also has the benefit of potentially facilitating the coordinated response to a stress since many stresses require more than one response for tolerance to occur.

In addition, the observation that many abiotic stresses can impinge upon a plant simultaneously has created much interest in investigating the possibility of generating plants with catch-all alterations involving the signaling pathways and their early responses that are common to several abiotic stresses. This could be done by altering levels or patterns of expression of higher level transcription factors involved in the early responses to stresses, as has been described to some extent in Arabidopsis. Although increased tolerance was associated with reduced growth in unstressed conditions, placing such genes behind a strong stressinducible promoter reduced this growth inhibition in unstressed conditions. However, such an approach has not been extensively tested in grasses, although field trials of wheat expressing DREB1A have commenced in Mexico.

Using the same rationale of trying to prevent damage from occurring in the first place, increased tolerance to NaCl was observed in plants in which expression of a Na^+ transporter, HKT1, was reduced by antisense. This was associated with a decrease in net Na^+ influx and root Na^+ concentration. The potential of this manipulation to increase Na^+ tolerance in wheat is significant, especially given the clear empirical observations, such as by Munns et al., of Na^+ exclusion being proportional to Na^+ tolerance. Unfortunately, some of the alterations described above appear to come with a growth and/or yield penalty in conditions of reduced stress. A strategy to avoid this outcome would be to drive expression of these genes in response to stress by an inducible promoter, as has been done for controlling expression of a gene encoding a key Pro-synthesising enzyme in rice. This has successfully removed inhibition of growth in low-stress conditions. Such an approach should now be attempted for manipulations in wheat.

Physiological knowledge of the processes of abiotic stress tolerance in grasses is still developing, and it is clear that significantly more effort needs to be invested to both complement and guide both breeding and genetic manipulation programmes. It would be desirable that future work exploit further the synergies to be gained by the interfacing of physiological and molecular/genetic research.

Transformation Methods in Transgenic Plants

The use of genetic engineering techniques for introducing genes responsible for agronomic characteristics may facilitate the development of new varieties. In fact, the last two decades have seen significant developments in plant transformation technology, and many transgenic crop plants have now been released for commercial production. Advances in the technology have been due to the development of a range of *Agrobacterium*-mediated and direct DNA delivery techniques, along with appropriate tissue culture techniques for regenerating whole plants from plant cells or tissues in a large number of species. To obtain a transgenic plant, methods of introducing the exogenous DNA into the vegetable tissue and regenerating the whole plant from such genetically transformed tissue are required.

Depending upon the species to be transformed, various types of tissue have been used. The meristematic tissue has been preferably employed in various transformation processes, primarily due to the ease of regeneration of a plant from this type of tissue. Several methods have been developed for inserting genetic information into plant cells, such as Agrobacterium-mediated systems, direct DNA uptake into protoplasts and particle bombardment (biolistic). However, although it was assumed that advances in the methodologies for gene delivery and plant tissue culture would make plant transformation routine for most important crops, some species have proved very difficult to transform.

Several methodologies are available to transform almost all important crops, however, some systems are still not routine and do not achieve the frequency of transformation required for obtaining genetically modified plants containing useful traits, such as abiotic stress tolerance.

Particle Bombardment and Plant Transformation

Since the development of the biolistic process for the direct introduction of genes into vegetable cells at the end of the 1980s, a great number of transgenic plants of several species have been obtained, including those species that were considered difficult to transform by other methods. This is due to the fact that it has become possible to introduce and express exogenous genes in any kind of vegetable tissue. Thus, any type of tissue having the potential to regenerate a completely fertile plant is suitable for transformation.

The biolistic process has also been called the microprojectile bombardment method, "gene gun" method, and particle acceleration method, among others. Different systems have been developed and constructed which are capable of accelerating microparticles, coated with nucleic acid sequences, at speeds exceeding 1,500 km/h.

All these systems are based on the generation of a shock wave with enough energy to displace microparticles coated with DNA. The shock wave can be generated by a chemical explosion (dry powder gun), a discharge of helium gas under high pressure, or vaporization of a drop of water through an electric discharge at high voltage and low capacitance or at low voltage and high capacitance.

Those systems that use helium gas under high pressure and electric discharge have shown a broad range of uses. The accelerated particles penetrate the cellular wall and membrane in a non-lethal way, locating themselves randomly in the cellular organells. Then the DNA is dissociated from the microparticles by the action of the cellular liquid, and the process of integrating the exogenous DNA in the genome of the organism to be modified takes place. In spite of the efficiency and universality of utilisation of the biolistic process, it requires the optimisation of various biological parameters, related to tissue culture, allowing regeneration of transgenic plants from genetically modified tissues.

Several protocols have been developed to *de novo* regenerate the most important crops. However, genetic modification of some of them is still not routine, with low frequency of transformation, or limited to some specific varieties. Consequently, the apical region of embryonic axes is an obvious target for the development of systems based on the bombardment of meristematic cells.

Based on the introduction of genes into the apical meristematic cells, by particle bombardment some researchers developed an efficient and reproducible system to achieve routinely transgenic plants of several species, such as soybean, dry bean and cotton. The ability to recover a high number of fertile transgenic plants may facilitate the detection of the "key event(s)" that could be integrated into plant breeding programmes for the development of transgenic crops that will be introduced into the market.

Molecular analyses and progeny tests of several generations of transgenic lines revealed the presence of a small number of integrated copies of the foreign genes and segregation in a Medelian fashion in most of them. This is extremely important, in order to accelerate the introduction of these plants in a breeding programme as well as the production of transgenic commercial varieties. Embrapa's groups and others are utilising these systems in order to introduce genes for abiotic stress in important crops such as bean and soybean, namely drought and cold tolerance.

Dry Bean Transformation

The common bean (*Phaseolus vulgaris* L.) is one of the most important food legumes in the developing world. In Latin America, beans are a very important source of protein and calories. However, the production growth rates have been declining in Brazil and the Andean regions, mainly due to diseases, insects, nutritional deficiencies, and drought. Consequently, there is considerable interest in the introduction of agronomically useful traits into beans by breeding and genetic engineering.

Several methods have been developed for inserting genetic information into plant cells, such as Agrobacterium-mediated systems, direct DNA uptake into protoplasts and particle bombardment. Early efforts to transform the bean demonstrated its susceptibility to *Agrobacterium*, and some transgenic tissues such as callus, leaves, meristems, cotyledon and hypocotyl have been achieved. Mariotti et al. say that the production of transgenic bean plants through the utilisation of the *Agrobacterium* system. However, there was no molecular evidence for genetic transformation or progeny analysis. Transient gene expression using either electroporation or PEG-mediated protoplast transformation was demonstrated. Dillen et al. demonstrated the applicability of electroporation of intact tissue to introduce and express the *gus* gene in bean embryonic axes. Nevertheless, no transgenic bean plants were obtained due to the difficulties of regenerating plants.

The applicability of particle bombardment to introduce and transiently express genes into apical meristematic cells of dry beans was demonstrated in the early 1990s. However, it was not known how deep the particles should penetrate in order to reach the cells that could generate transgenic plants. The different parts of the meristem have been divided in layers (L1, L2 and L3). Layer L1 is the most external and forms the epidermis of the differentiated regions. Layers L2 and L3 divide preferentially in the anticlinal and periclinal planes to form the organs. Several experiments have demonstrated that the differentiated *de novo* shoots originate from the sub-epidermal layers (L2 and L3) of the apical meristem. However, the L1 layer could participate in their formation. These shoots are formed in the peripheral regions of the apical meristem.

The bombardment of *P. vulgaris* meristematic cells showed that it was possible to efficiently reach these layers, demonstrating that it would be possible to achieve transgenic plants. In 1993, Russel et al. were able to achieve transgenic navy bean plants using an electrical particle acceleration device. It was the first report of *P. vulgaris* transformation, presenting molecular evidences and progeny. However, the frequency of transgenic plants obtained was much lower (0.03%) and variety-limited.

In addition, their tissue culture protocol was time consuming, involving several temperature treatments and medium transfers of the bombarded embryos before recovery of transgenic shoots. Kim and Minamikawa achieved transformation by bombarding embryonic axes, obtaining stable transformed bean plants. The researchers developed an efficient and reproducible system to achieve routinely transgenic bean plants.

The bean transformation system was also based upon the development of a tissue culture protocol of multiple shoot induction, shoot elongation and rooting. The average frequency of transformation was 0.9%. It has been assumed that the use of particle bombardment to introduce genes into apical meristematic cells can be extended to any genotype or variety. The morphology of the explants utilised during the bombardment process may greatly influence the successful recuperation achievement of transgenic bean plants.

In some cultivars, the embryonic axes revealed the apical meristematic region partially exposed, whereas only the central region could be visualised. The number of meristematic cells that could be reached by the microparticles coated with DNA will be drastically reduced. Consequently, the efficiency of transformation could also be reduced. Several experiments have shown that *de novo* shoot differentiation in embryos of bean cultivated on cytokinins appeared in the peripheral layers of the meristematic ring. Thus, based on these concepts, cultivars with a non-exposed apical meristematic region are not suitable for particle bombardment transformation, considering the fact that removal of the leaf primordia is not practical.

Mature embryos were transformed in order to transiently express a methionine-rich albumin gene from Brazil nut (*be2s1* gene), which could be detected by Western blot and ELISA, twenty-four hours later. Recently, they stably transformed bean with the *be2s1* gene in order to improve the methionine content in the seeds. In two of the five transgenic lines, the methionine content was significantly increased by 14% and 23% over the non-transgenic plants.

Recently, obtained transgenic plants containing the bar gene that confers resistance to the herbicide glufosinate ammonium. The bar gene was found to confer strong resistance in transgenic bean to the herbicide in the greenhouse. A field trial was carried out and one line presented high tolerance up to 400 g/ha of glufosinate ammonium. In order to achieve bean plants resistant to the bean golden mosaic geminivirus, which is responsible for one of the most important diseases in this crop in Latin America, plants containing the mutated *AC1* viral gene were obtained. The two lines that presented immunity against the virus. These lines were introduced into the breeding programme from the Center of Rice and Bean of Embrapa

The bean transformation system developed in lab is now available to introduce genes for abiotic stress tolerance. Indeed, other groups in Mexico and Brazil are using this technology to introduce genes for tolerance to cold, drought and shade environment.

Soybean Transformation

Soybean transformation involved the use of the *Agrobacterium* system to obtain plants regenerated from cotyledonary nodes and the biolistic system, which is based on acceleration of microparticles coated with DNA into the apical meristematic region of embryonic axes. Since then, several groups have been actively involved in improving these systems by manipulating somatic embryogenesis and shoot morphogenesis.

In addition, an efficient selectable marker that would allow the regeneration of only transgenic material and increase the frequency of recovery of fertile transgenic plants is not available.

One important constraint in the transformation system based on the bombardment of meristematic tissue of embryonic axes was the difficulty of achieving efficient selection of transformed cells. In this case, only a few cells from the meristem are transformed, while the others, including those that are in contact with the culture medium and the selective agent, are not transformed.

Antibiotics such as kanamicin have been used as selective agents, associated with the *nptII* gene. However, this selection is not efficient. A soybean transformation technology that would allow the selection of transgenic meristematic cells and regeneration of fertile plants should facilitate the production of transgenic plants with improved agricultural characteristics.

To obtain transgenic soybean plants from the apical region of embryonic axes, there are two essential requirements: 1) introduction of exogenous genes with high frequencies into the cells of the apical regions, and integration of exogenous genes into the vegetable genome, and 2) regeneration and production of fertile transgenic plants from the resulting transformed cells.

With the development of the biolistic process, the first requirement was achieved. However, one problem remained: it was difficult to select the transformed apical meristematic cells. To solve this problem, which is one of the most important limiting factors for achieving transgenic soybean with a high frequency, they developed a novel system to select transgenic meristematic cells. The system is based on the use of the selective agent imazapyr, a herbicidal molecule of the imidazolinone class, capable of systemically translocating and concentrating in the apical meristematic region of the plant. The mechanism of action of imazapyr is the inhibition of the enzymatic activity of acetohydroxyacid synthase, which catalyses the initial step in the biosynthesis of isoleucine, leucine and valine.

The *ahas* selectable marker gene, previously isolated from Arabidopsis thaliana, contains a mutation at position 653 bp resulting in a serine to asparagine substitution that confers imidazolinone-specific resistance. This selectable marker system, combined with an improved multiple shooting induction protocol, resulted in a significant increase in the recovery of fertile, transgenic material compared with the standard soybean transformation protocol (0.1%).

Basically, mature seeds are surface sterilised and soaked in distilled water for 18-20 h. The embryonic axes are excised from seeds and the apical meristems are exposed by removing the primary leaves, and positioned for the bombardment. After bombardment, the embryonic axes are transferred and completely immersed in the induction medium, for 16 h in the dark. After this period, the explants are transferred to MS medium containing 500 nM imazapyr. As soon as the embryonic axesderived shoots are 2 to 3 cm in length, they are analysed for the presence of foreign genes and individually transferred to a plastic pot and acclimated. This technique is variety independent. It was possible to transform several commercial cultivars, such as BR-16, Doko RC, BR-91, Guarany and Conquista. One line, from the cultivar BR-16, was tested under field conditions and presented very high tolerance to the herbicide imazapyr.

This new technology should have significant applications for the introduction of genes for agronomic characteristics such as biotic and abiotic stress tolerance. In fact, this soybean transformation system will be used for the introduction and expression of the *DREB1A* gene identified in *Arabidopsis thaliana* and involved in the response of the plants to abiotic stress.

Cotton Transformation

Cotton (*Gossypium hirsutum* L.) is considered a high value agricultural product, adapted for the industrial production of vegetable oil and mainly textile fiber. The sharp increase in cotton consumption in the form of greater variety of products justifies the need to optimise the production and quality of this crop. Genetic engineering can help improve cotton breeding programmes. The first reports of cotton transformation, using Agrobacterium tumefaciens, were applicable to only a few varieties. Most varieties of commercial interest proved difficult or impossible to regenerate into plants from the obligatory callus phase. Genes introduced into regenerable varieties could be bred into other lines but the process is lengthy and prone to somaclonal variation.

Most of the varieties transformed by Agrobacterium systems are Coker and Coker-derived. In this regard, looking for an efficient system to transfer genes to elite cultivars, avoiding the need of many years of crossing and selection to reach new varieties.

Based on their experience with soybean, McCabe and Martinell developed a variety-independent protocol for the production of transgenic cotton. However, its efficiency was very low, from 0.027 to 0.2%, which is not sufficient for practical purposes. Stable integration and expression of the genes have been achieved with both *Agrobacterium* and the biolistic process, however when only the epidermal cortex cell layer of a plant is transformed, the progeny will not inherit the transgene.

Germline transformants, on the other hand, pass on the transgenes to their progeny and are suitable for conventional commercial cultivation. The frequency of cotton transformation has been relatively high (0.2 to 1%); nevertheless, the frequency of germline transformation has been very low. This is the most important constraint for the achievement of a variety-independent system with high frequency of cotton transformation. They successfully adapted the selection system developed for soybean to increase the cotton germline frequency of transformation.

Indeed, the frequency of germline transformation was raised to 0.5%, using the herbicide imazapyr to select the shoots and plantlets originated from the apical meristematic tissue bombarded with the ahas gene. Seeds of cotton, varieties CNPA-7MH and 7H, are surface sterilised and soaked in distilled water for 16 h. The water is removed and the seeds are incubated in the dark for an additional period of 16 h. Then, the embryonic axes are excised from seeds and the apical meristems are exposed by removing the primary leaves, and positioned for the bombardment (Figure 1).

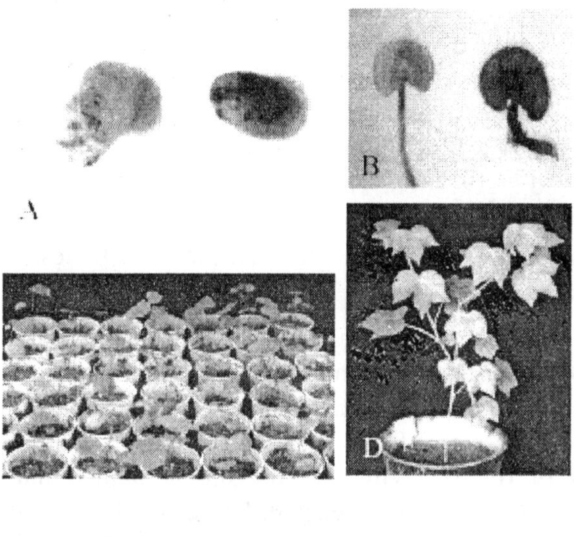

Figure 1. Cotton transformation system. A) Embryonic axes expressing the gus *(uidA) gene 3 days after the bombardment. B) Transgenic (right) and non-transgenic anthers expressing the* gus *gene. C) Acclimated transgenic cotton plantlets (R0). D) A transgenic cotton plant in the greenhouse.*

After bombardment the embryonic axes are transferred to MS medium containing 900 nM imazapyr, 0.1% activated charcoal,

3 mg/L BAP and 3% glucose. As soon as the embryonic axes-derived shoots are 2 to 3 cm in length, they are analysed for the presence of foreign genes and individually transferred to a plastic pot and acclimated (Figure 1). Most of the elongated shoots showed the presence of foreign genes. The developed process should allow the introduction and expression of different genes in elite varieties of commercial cotton, with a frequency of germline transformation superior to those protocols described so far.

Crop improvement through genetic engineering has become a reality. For some crops, transformation is already routine. However, for other crops, despite the possibility of obtaining transgenic plants, the frequency is still not sufficient for practical purposes. In experience, we have seen that for each 20 transgenic plants obtained, only one has the potential to be introduced in a breeding programme to generate a commercial variety.

Consequently, a transformation technology has to be very efficient in order to allow the selection of events with a desirable expression level, without severe somaclonal variation and unexpected segregation. This is particularly true for introducing genes for abiotic stress tolerance, since either the number of candidate genes or that of individual plants to be tested should be large.

In addition, research in plant transformation is concentrating now not so much on the introduction of DNA into plant cells, but rather on the problems associated with stable integration and reliable expression of the DNA once it has been integrated. Studies of the behaviour of transgenic plants under field conditions need to be conducted. These studies should assess the interactions of the transgenic plants with other plants in the agricultural and natural environment. Moreover, the stability of foreign genes and factors related to their interaction with the complex physiology of these plants submitted to natural stress in tropical areas should be evaluated.

9

Plant Tissue Culture and Regeneration

Most methods of plant transformation applied to genetic modification (GM) crops require that a whole plant is regenerated from isolated plant cells or tissue which have been genetically transformed. This regeneration is conducted *in vitro* so that the environment and growth medium can be manipulated to ensure a high frequency of regeneration. In addition to a high frequency of regeneration, the regenerable cells must be accessible to gene transfer by whatever technique is chosen.

The primary aim is therefore to produce, as easily and as quickly as possible, a large number of regenerable cells that are accessible to gene transfer. The subsequent regeneration step is often the most difficult step in plant transformation studies. However, it is important to remember that a high frequency of regeneration does not necessarily correlate with high transformation efficiency.

In this chapter we can see some basic issues concerned with plant tissue culture *in vitro*, particularly as applied to plant transformation. We can also see that the basic culture types used for plant transformation and some of the techniques that can be used to regenerate whole transformed plants from transformed cells or tissue.

Plant Tissue Culture

Practically any plant transformation experiment relies at some point on tissue culture. There are some exceptions to this generalisation, but the ability to regenerate plants from isolated cells or tissues *in vitro* underpins most plant transformation systems.

Plasticity and Totipotency

Two concepts, plasticity and totipotency, are central to understanding plant cell culture and regeneration. Plants, due to their sessile nature and long life span, have developed a greater ability to endure extreme conditions and predation than have animals. Many of the processes involved in plant growth and development adapt to environmental conditions. This plasticity allows plants to alter their metabolism, growth and development to best suit their environment.

Particularly important aspects of this adaptation, as far as plant tissue culture and regeneration are concerned, are the abilities to initiate cell division from almost any tissue of the plant and to regenerate lost organs or undergo different developmental pathways in response to particular stimuli. When plant cells and tissues are cultured *in vitro* they generally exhibit a very high degree of plasticity, which allows one type of tissue or organ to be initiated from another type.

In this way, whole plants can be subsequently regenerated. This regeneration of whole organisms depends upon the concept that all plant cells can, given the correct stimuli, express the total genetic potential of the parent plant. This maintenance of genetic potential is called 'totipotency'. Plant cell culture and regeneration do, in fact, provide the most compelling evidence for totipotency. In practical terms though, identifying the culture conditions and stimuli required to manifest this totipotency can be extremely difficult and it is still a largely empirical process.

The culture environment

When cultured *in vitro*, all the needs, both chemical (Table 1) and physical, of the plant cells have to met by the culture vessel, the growth medium and the external environment (light, temperature, etc.). The growth medium has to supply all the essential mineral ions required for growth and development. In many cases, it must also supply additional organic supplements such as amino acids and vitamins.

Table 1. Some of the elements important for plant nutrition and their physiological function.

Element	Function
Nitrogen	Component of proteins, nucleic acids and some coenzymes. Element required in greatest amount
Potassium	Regulates osmotic potential, principal inorganic cation
Calcium	Cell wall synthesis, membrane function, cell signalling
Magnesium	Enzyme cofactor, component of chlorophyll
Phosphorus	Component of nucleic acids, energy transfer, component of intermediates in respiration and photosynthesis
Sulphur	Component of some amino acids (methionine, cysteine) and some cofactors
Chlorine	Required for photosynthesis
Iron	Electron transfer as a component of cytochromes
Manganese	Enzyme cofactor
Cobalt	Component of some vitamins
Copper	Enzyme cofactor, electron-transfer reactions
Zinc	Enzyme cofactor, chlorophyll biosynthesis
Molybdenum	Enzyme cofactor, component of nitrate reductase

Many plant cell cultures, as they are not photosynthetic, also require the addition of a fixed carbon source in the form of a sugar (most often sucrose). One other vital component that must also be supplied is water, the principal biological solvent. Physical factors, such as temperature, pH, the gaseous environment, light (quality and duration) and osmotic pressure, also have to be maintained within acceptable limits.

Culture Media

Culture media used for the *in vitro* cultivation of plant cells are composed of three basic components:

(1) essential elements, or mineral ions, supplied as a complex mixture of salts;

(2) an organic supplement supplying vitamins and/or amino acids; and

(3) a source of fixed carbon; usually supplied as the sugar sucrose.

For practical purposes, the essential elements are further divided into the following categories:

(1) macroelements (or macronutrients);

(2) microelements (or micronutrients); and

(3) an iron source.

Complete, plant cell culture medium is usually made by combining several different components, as outlined in Table 2.

Media Components

It is useful to briefly consider some of the individual components of the stock solutions.

Macroelements

As is implied by the name, the stock solution supplies those elements required in large amounts for plant growth and development. Nitrogen, phosphorus, potassium, magnesium, calcium and sulphur are usually regarded as macroelements.

These elements usually comprise at least 0.1% of the dry weight of plants.

Table 2. Composition of a typical plant culture medium.

Essential element	*Concentration in stock solution (mg/l)*	*Concentration in medium (mg/l)*
Macroelements		
NH_4NO_3	33000	1650
KNO_3	38000	1900
$CaCl_2 \times 2H_2O$	8800	440
$MgSO_4 \times 7H_2O$	7400	370
KH_2PO_4	3400	170
Microelements		
KI	166	0.83
H_3BO_3	1240	6.2
$MnSO_4 \times 4H_2O$	4460	22.3
$ZnSO_4 \times 7H_2O$	1720	8.6
$Na_2MoO_4 \times 2H_2O$	50	0.25
$CuSO_4 \times 5H_2O$	5	0.025
$CoCl_2 \times 6H_2O$	5	0.025
Iron source		
$FeSO_4 \times 7H_2O$	5560	27.8
$Na_2EDTA \times 2H_2O$	7460	37.3
Organic supplement		
Myoinositol	20000	100
Nicotinic acid	100	0.5
Pyridoxine-HCl	100	0.5
Thiamine-HCl	100	0.5
Glycine	400	2
Carbon source		
Sucrose	Added as solid	30000

Nitrogen is most commonly supplied as a mixture of nitrate ions (from the KNO_3) and ammonium ions (from the NH_4NO_3).

Theoretically, there is an advantage in supplying nitrogen in the form of ammonium ions, as nitrogen must be in the reduced form to be incorporated into macromolecules. Nitrate ions therefore need to be reduced before incorporation. However, at high concentrations, ammonium ions can be toxic to plant cell cultures and uptake of ammonium ions from the medium causes acidification of the medium.

In order to use ammonium ions as the sole nitrogen source, the medium needs to be buffered. High concentrations of ammonium ions can also cause culture problems by increasing the frequency of vitrification. Using a mixture of nitrate and ammonium ions has the advantage of weakly buffering the medium as the uptake of nitrate ions causes OH^- ions to be excreted. Phosphorus is usually supplied as the phosphate ion of ammonium, sodium or potassium salts. High concentrations of phosphate can lead to the precipitation of medium elements as insoluble phosphates.

Microelements

These elements are required in trace amounts for plant growth and development, and have many and diverse roles. Manganese, iodine, copper, cobalt, boron, molybdenum, iron and zinc usually comprise the microelements, although other elements such as nickel and aluminium are frequently found in some formulations. Iron is usually added as iron sulphate, although iron citrate can also be used. Ethylenediaminetetraacetic acid (EDTA) is usually used in conjunction with the iron sulphate. The EDTA complexes with the iron so as to allow the slow and continuous release of iron into the medium. Uncomplexed iron can precipitate out of the medium as ferric oxide.

Organic supplements

Only two vitamins, thiamine (vitamin B_1) and myoinositol (considered a B vitamin) are considered essential for the culture of plant cells *in vitro*. However, other vitamins are often added to plant cell culture media for historical reasons. Amino acids are also commonly included in the organic supplement. The most frequently used is glycine (arginine, asparagine, aspartic acid, alanine, glutamic acid, glutamine and proline are also used), but in many cases its inclusion is not essential. Amino acids provide a source of reduced nitrogen and, like ammonium ions, uptake causes acidification of the medium. Casein hydrolysate can be used as a relatively cheap source of a mix of amino acids.

Carbon source

Sucrose is cheap, easily available, readily assimilated and relatively stable and is therefore the most commonly used carbon source. Other carbohydrates (such as glucose, maltose, galactose and sorbitol) can also be used, and in specialised circumstances may prove superior to sucrose.

Gelling agents

Media for plant cell culture *in vitro* can be used in either liquid or 'solid' forms, depending on the type of culture being grown. For any culture types that require the plant cells or tissues to be grown on the surface of the medium, it must be solidified. Agar, produced from seaweed, is the most common type of gelling agent, and is ideal for routine applications. However, because it is a natural product, the agar quality can vary from supplier to supplier and from batch to batch. For more demanding applications, a range of purer gelling agents are available. Purified agar or agarose can be used, as can a variety of gellan gums.

Growth Regulators

Plant growth regulators are the critical media components in determining the developmental pathway of the plant cells. The plant growth regulators used most commonly are plant hormones or their synthetic analogues.

There are five main classes of plant growth regulator used in plant cell culture, namely:

(1) auxins;
(2) cytokinins;
(3) gibberellins;
(4) abscisic acid;
(5) ethylene.

Auxins

Auxins promote both cell division and cell growth The most important naturally occurring auxin is IAA (indole-3-acetic acid), but its use in plant cell culture media is limited because it is unstable to both heat and light. Occasionally, amino acid conjugates of IAA (such as indole-acetyl-L-alanine and indole-acetyl-L-glycine), which are more stable, are used to partially alleviate the problems associated with the use of IAA.

Table 3. Commonly used auxins, their abbreviation and chemical name

Abbreviation/name	Chemical name
2,4-D	2,4-dichlorophenoxyacetic acid
2,4,5-T	2,4,5-trichlorophenoxyacetic acid
Dicamba	2-methoxy-3,6-dichlorobenzoic acid
IAA	Indole-3-acetic acid
IBA	Indole-3-butyric acid
MCPA	2-methyl-4-chlorophenoxyacetic acid
NAA	1-naphthylacetic acid
NOA	2-naphthyloxyacetic acid
Picloram	4-amino-2,5,6-trichloropicolinic acid

It is more common, though, to use stable chemical analogues of IAA as a source of auxin in plant cell culture media. 2,4-Dichlorophenoxyacetic acid (2,4-D) is the most commonly used auxin and is extremely effective in most circumstances. Other auxins are available (Table 3), and some may be more effective or 'potent' than 2,4-D in some instances.

Cytokinins

Cytokinins promote cell division. Naturally occurring cytokinins are a large group of structurally related (they are purine derivatives) compounds. Of the naturally occurring cytokinins, two have some use in plant tissue culture media (Table 4). These are zeatin and 2iP (2-isopentyl adenine). Their use is not widespread as they are expensive (particularly zeatin) and relatively unstable.

Table 4. Commonly used cytokinins, their abbreviation and chemical name

Abbreviation/name	*Chemical name*
BAP	6-benzylaminopurine
2iP (IPA)	[N^6-(2-isopentyl)adenine]
Kinetin	6-furfurylaminopurine
Thidiazuron	1-phenyl-3-(1,2,3-thiadiazol-5-yl)urea
Zeatin	4-hydroxy-3-methyl-trans-2-butenylaminopurine

The synthetic analogues, kinetin and BAP (benzylaminopurine), are therefore used more frequently. Non-purine-based chemicals, such as substituted phenylureas, are also used as cytokinins in plant cell culture media. These substituted phenylureas can also substitute for auxin in some culture systems.

Gibberellins

There are numerous, naturally occurring, structurally related compounds termed 'gibberellins'. They are involved in regulating cell elongation, and are agronomically important in determining plant height and fruit-set. Only a few of the gibberellins are used in plant tissue culture media, GA_3 being the most common.

Abscisic acid

Abscisic acid (ABA) inhibits cell division. It is most commonly used in plant tissue culture to promote distinct developmental pathways such as somatic embryogenesis.

Ethylene

Ethylene is a gaseous, naturally occurring, plant growth regulator most commonly associated with controlling fruit ripening in climacteric fruits, and its use in plant tissue culture is not widespread. It does, though, present a particular problem for plant tissue culture. Some plant cell cultures produce ethylene, which, if it builds up sufficiently, can inhibit the growth and development of the culture. The type of culture vessel used and its means of closure affect the gaseous exchange between the culture vessel and the outside atmosphere and thus the levels of ethylene present in the culture.

Generalisations about plant growth regulators and their use in plant cell culture media have been developed from initial observations made in the 1950s. There is, however, some considerable difficulty in predicting the effects of plant growth regulators: this is because of the great differences in culture response between species, cultivars and even plants of the same cultivar grown under different conditions.

However, some principles do hold true and have become the paradigm on which most plant tissue culture regimes are based. Auxins and cytokinins are the most widely used plant growth regulators in plant tissue culture and are usually used together, the ratio of the auxin to the cytokinin determining the type of culture established or regenerated (Figure 1). A high auxin to cytokinin ratio generally favours root formation, whereas a high cytokinin to auxin ratio favours shoot formation. An intermediate ratio favours callus production.

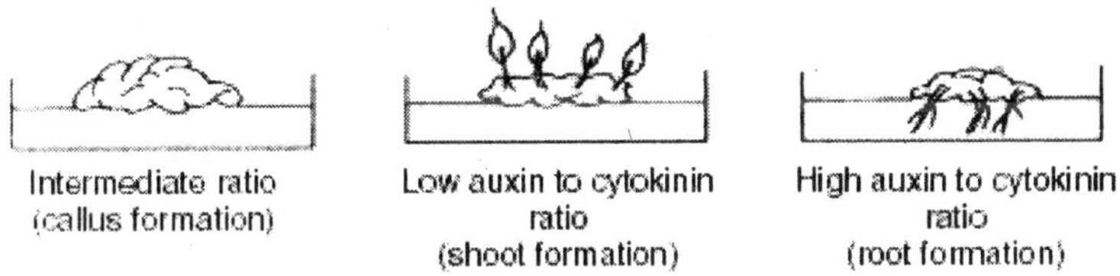

Figure 1. The effect of different ratios of auxin to cytokinin on the growth and morphogenesis of callus.

Culture Types

Cultures are generally initiated from sterile pieces of a whole plant. These pieces are termed 'explants', and may consist of pieces of organs, such as leaves or roots, or may be specific cell types, such as pollen or endosperm. Many features of the explant are known to affect the efficiency of culture initiation. Generally, younger, more rapidly growing tissue (or tissue at an early stage of development) is most effective.

Callus

Explants, when cultured on the appropriate medium, usually with both an auxin and a cytokinin, can give rise to an unorganised, growing and dividing mass of cells. It is thought that any plant tissue can be used as an explant, if the correct conditions are found. In culture, this proliferation can be maintained more or less indefinitely, provided that the callus is subcultured on to fresh medium periodically. During callus formation there is some degree of dedifferentiation (i.e. the changes that occur during development and specialisation are, to some extent, reversed), both in morphology (callus is usually composed of unspecialised parenchyma cells) and metabolism. One major consequence of this dedifferentiation is that most plant cultures lose the ability to photosynthesise.

This has important consequences for the culture of callus tissue, as the metabolic profile will probably not match that of the donor plant. This necessitates the addition of other components-such as vitamins and, most importantly, a carbon source-to the culture medium, in addition to the usual mineral nutrients.

Callus culture is often performed in the dark (the lack of photosynthetic capability being no drawback) as light can encourage differentiation of the callus.

During long-term culture, the culture may lose the requirement for auxin and/or cytokinin. This process, known as 'habituation', is common in callus cultures from some plant species (such as sugar beet). Callus cultures are extremely important in plant biotechnology. Manipulation of the auxin to cytokinin ratio in the medium can lead to the development of shoots, roots or somatic embryos from which whole plants can subsequently be produced. Callus cultures can also be used to initiate cell suspensions, which are used in a variety of ways in plant transformation studies.

Cell-suspension Cultures

Callus cultures, broadly speaking, fall into one of two categories: compact or friable. In compact callus the cells are densely aggregated, whereas in friable callus the cells are only loosely associated with each other and the callus becomes soft and breaks apart easily. Friable callus provides the inoculum to form cell-suspension cultures. Explants from some plant species or particular cell types tend not to form friable callus, making cell-suspension initiation a difficult task. The friability of callus can sometimes be improved by manipulating the medium components or by repeated subculturing. The friability of the callus can also sometimes be improved by culturing it on 'semi-solid' medium (medium with a low concentration of gelling agent). When friable callus is placed into a liquid medium (usually the same composition as the solid medium used for the callus culture) and then agitated, single cells and/or small clumps of cells are released into the medium. Under the correct conditions, these released cells continue to grow and divide, eventually producing a cell-suspension culture.

A relatively large inoculum should be used when initiating cell suspensions so that the released cell numbers build up quickly. The inoculum should not be too large though, as toxic products released from damaged or stressed cells can build up to lethal levels. Large cell clumps can be removed during subculture of the cell suspension.

Cell suspensions can be maintained relatively simply as batch cultures in conical flasks. They are continually cultured by repeated subculturing into fresh medium. This results in dilution of the suspension and the initiation of another batch growth cycle. The degree of dilution during subculture should be determined empirically for each culture. Too great a degree of dilution will result in a greatly extended lag period or, in extreme cases, death of the transferred cells.

After subculture, the cells divide and the biomass of the culture increases in a characteristic fashion, until nutrients in the medium are exhausted and/or toxic by-products build up to inhibitory levels-this is called the 'stationary phase'. If cells are left in the stationary phase for too long, they will die and the culture will be lost. Therefore, cells should be transferred as they enter the stationary phase. It is therefore important that the batch growth-cycle parameters are determined for each cell-suspension culture.

Protoplasts

Protoplasts are plant cells with the cell wall removed. Protoplasts are most commonly isolated from either leaf mesophyll cells or cell suspensions, although other sources can be used to advantage. Two general approaches to removing the cell wall (a difficult task without damaging the protoplast) can be taken-mechanical or enzymatic isolation.

Mechanical isolation, although possible, often results in low yields, poor quality and poor performance in culture due to substances released from damaged cells. Enzymatic isolation is usually carried out in a simple salt solution with a high osmoticum, plus the cell wall degrading enzymes. It is usual to use a mix of both cellulase and pectinase enzymes, which must be of high quality and purity.

Protoplasts are fragile and easily damaged, and therefore must be cultured carefully. Liquid medium is not agitated and a high osmotic potential is maintained, at least in the initial stages. The liquid medium must be shallow enough to allow aeration in the absence of agitation.

Protoplasts can be plated out on to solid medium and callus produced. Whole plants can be regenerated by organogenesis or somatic embryogenesis from this callus. Protoplasts are ideal targets for transformation by a variety of means.

Root Cultures

Root cultures can be established *in vitro* from explants of the root tip of either primary or lateral roots and can be cultured on fairly simple media. The growth of roots *in vitro* is potentially unlimited, as roots are indeterminate organs. Although the establishment of root cultures was one of the first achievements of modern plant tissue culture, they are not widely used in plant transformation studies.

Shoot Tip and Meristem Culture

The tips of shoots (which contain the shoot apical meristem) can be cultured *in vitro*, producing clumps of shoots from either axillary or adventitious buds. This method can be used for clonal propagation.

Shoot meristem cultures are potential alternatives to the more commonly used methods for cereal regeneration as they are less genotype-dependent and more efficient (seedlings can be used as donor material).

Embryo Culture

Embryos can be used as explants to generate callus cultures or somatic embryos. Both immature and mature embryos can be used as explants. Immature, embryo-derived embryogenic callus is the most popular method of monocot plant regeneration.

Microspore Culture

Haploid tissue can be cultured *in vitro* by using pollen or anthers as an explant. Pollen contains the male gametophyte, which is termed the 'microspore'. Both callus and embryos can be produced from pollen. Two main approaches can be taken to produce *in vitro* cultures from haploid tissue. The first method depends on using the anther as the explant. Anthers (somatic tissue that surrounds and contains the pollen) can be cultured on solid medium (agar should not be used to solidify the medium as it contains inhibitory substances).

Pollen-derived embryos are subsequently produced via dehiscence of the mature anthers. The dehiscence of the anther depends both on its isolation at the correct stage and on the correct culture conditions. In some species, the reliance on natural dehiscence can be circumvented by cutting the wall of the anther, although this does, of course, take a considerable amount of time. Anthers can also be cultured in liquid medium, and pollen released from the anthers can be induced to form embryos, although the effi-ciency of plant regeneration is often very low. Immature pollen can also be extracted from developing anthers and cultured directly, although this is a very time-consuming process.

Both methods have advantages and disadvantages. Some beneficial effects to the culture are observed when anthers are used as the explant material. There is, however, the danger that some of the embryos produced from anther culture will originate from the somatic anther tissue rather than the haploid microspore cells. If isolated pollen is used there is no danger of mixed embryo formation, but the efficiency is low and the proccss is time-consuming.

In microspore culture, the condition of the donor plant is of critical importance, as is the timing of isolation. Pretreatments, such as a cold treatment, are often found to increase the efficiency. These pretreatments can be applied before culture, or, in some species, after placing the anthers in culture. Plant species can be divided into two groups, depending on whether they require the addition of plant growth regulators to the medium for pollen/anther culture; those that do also often require organic supplements, e.g. amino acids.

Many of the cereals (rice, wheat, barley and maize) require medium supplemented with plant growth regulators for pollen/ anther culture. Regeneration from microspore explants can be obtained by direct embryogenesis, or via a callus stage and subsequent embryogenesis. Haploid tissue cultures can also be initiated from the female gametophyte (the ovule). In some cases, this is a more efficient method than using pollen or anthers.

The ploidy of the plants obtained from haploid cultures may not be haploid. This can be a consequence of chromosome doubling during the culture period. Chromosome doubling (which often has to be induced by treatment with chemicals such as colchicine) may be an advantage, as in many cases haploid plants are not the desired outcome of regeneration from haploid tissues. Such plants are often referred to as 'di-haploids', because they contain two copies of the same haploid genome.

Plant Regeneration

Having looked at the main types of plant culture that can be established *in vitro,* can now look at how whole plants can be regenerated from these cultures. In broad terms, two methods of plant regeneration are widely used in plant transformation studies, i.e. somatic embryogenesis and organogenesis.

Somatic Embryogenesis

In somatic (asexual) embryogenesis, embryo-like structures, which can develop into whole plants in a way analogous to zygotic embryos, are formed from somatic tissues (Figure 2).

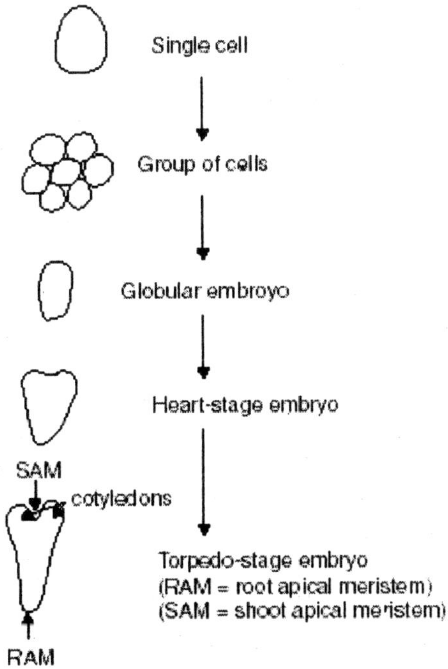

Figure 2. A schematic representation of the sequential stages of somatic embryo development.

These somatic embryos can be produced either directly or indirectly. In direct somatic embryogenesis, the embryo is formed directly from a cell or small group of cells without the production of an intervening callus. Though common from some tissues (usually reproductive tissues such as the nucellus, styles or pollen), direct somatic embryogenesis is generally rare in comparison with indirect somatic embryogenesis. In indirect somatic embryogenesis, callus is first produced from the explant. Embryos can then be produced from the callus tissue or from a cell suspension produced from that callus.

Somatic embryogenesis usually proceeds in two distinct stages. In the initial stage (embryo initiation), a high concentration of 2,4-D is used. In the second stage (embryo production) embryos are produced in a medium with no or very low levels of 2,4-D. In many systems it has been found that somatic embryogenesis is improved by supplying a source of reduced nitrogen, such as specific amino acids or casein hydrolysate.

Organogenesis

Somatic embryogenesis relies on plant regeneration through a process analogous to zygotic embryo germination. Organogenesis relies on the production of organs, either directly from an explant or from a callus culture. There are three methods of plant regeneration via organogenesis.

The first two methods depend on adventitious organs arising either from a callus culture or directly from an explant (Figure 3).

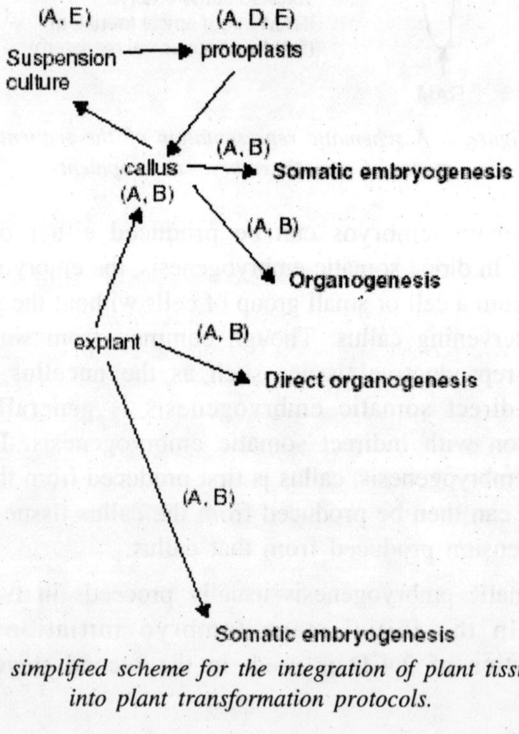

Figure 3. A simplified scheme for the integration of plant tissue culture into plant transformation protocols.

Alternatively, axillary bud formation and growth can also be used to regenerate whole plants from some types of tissue culture. Organogenesis relies on the inherent plasticity of plant tissues, and is regulated by altering the components of the medium. In particular, it is the auxin to cytokinin ratio of the medium that determines which developmental pathway the regenerating tissue will take. It is usual to induce shoot formation by increasing the cytokinin to auxin ratio of the culture medium. These shoots can then be rooted relatively simply.

Organogenesis in Tobacco

Organogenesis from tobacco pith callus is the classical example of how varying plant growth regulator regimes can be used to manipulate the pattern of regeneration from plant tissue cultures. When cultured on a medium containing both auxin and cytokinin, callus will proliferate. If the auxin to cytokinin ratio is increased, adventitious roots will form from the callus by organogenesis. It the auxin to cytokinin ratio is decreased adventitious shoots will be formed.

If the explants are cultured on medium containing only a cytokinin shoots can be produced directly. Tobacco plants can also be easily regenerated from tobacco leaf pieces. Leaves are cut into aproximately 1 cm squares with a sterile scalpel (avoiding large leaf veins and any damaged areas). The leaf pieces are then transferred (right side up) to gelled MS medium supplemented with 1mg l^{-1} BAP (a cytokinin) and 0.1mgl^{-1} NAA (an auxin). Over the next few weeks, callus forms on the explants, particularly around the cut surfaces. After 3 to 5 weeks shoots emerge directly from the explants or from callus derived from the explants. When these shoots are about 1cm long they can be cut at the base and placed on to solid MS medium without any plant growth regulators. The shoots will form roots and form plantlets that will grow in this medium and can subsequently be transferred to soil.

Tissue culture and plant regeneration are an integral part of most plant transformation strategies, and can often prove to be the most challenging aspect of a plant transformation protocol. Key to success in integrating plant tissue culture into plant transformation strategies is the realisation that a quick and efficient regeneration system must be developed. However, this system must also allow high transformation efficiencies from whichever transformation technique is adopted.

Not all regeneration protocols are compatible with all transformation techniques. Some crops may be amenable to a variety of regeneration and transformation strategies, others may currently only be amenable to one particular protocol. Advances are being made all the time, so it is impossible to say that a particular crop will never be regenerated by a particular protocol. However, some protocols, at least at the moment, are clearly more efficient than others. Regeneration from immature embryo-derived somatic embryos is, for example, the favoured method for regenerating monocot species.

10

Organic Plant Breeding

Organic producers try to achieve an optimal economic yield without exhausting natural resources in and around the production site while adhering to organic criteria. Instead of synthetic chemical fertilisers, veterinary drugs, pesticides, herbicides and growth hormones, more natural principles and methods are applied. Organic farmers strive for closed production cycles, natural self-regulation and agro-biodiversity.

Organic farming is more than just an environmentally-friendly production method, it is also a source of wholesome food. When asked why they believe that organic produce is better for them, consumers most often refer to the fact that organic crops are not treated with chemical pesticides, i.e. that they are free of residues. However, want to develop a vision of food quality that encompasses more than just the absence of residues. For this, the biodynamic vision on people and nature for inspiration. A vision on food quality will eventually lead to operational criteria for organic or biodynamic production methods, ensuring that these methods are environmentally-friendly and people-friendly.

The concept of plant vigour will be a crucial part of such criteria. Concerns about plant breeding must be addressed before organic agriculture can evolve further. Plant breeding systems specifically for organic production are few and far between, even though the organic sector is striving to achieve a completely organic production chain. Until now, organic farmers have often had little choice but to use conventionally bred varieties designed for conventional systems based on the use of synthetic chemical fertilisers and pesticides.

Despite these varieties' higher yield compared to old varieties—on organic farms, too—performance in organic systems could be vastly improved if varieties were bred for typically organic conditions. The organic sector has listed its wishes regarding plant characteristics (product quality) and methods to develop new varieties (production quality). These methods should take into account organic principles about breeding techniques as well as related socio-economic conditions and structures.

PARTIES IN ORGANIC PLANT BREEDING

Numerous people are involved in the production chain from breeder to consumer. There are three main parties: farmers (growers and producers), trade and industry (manufacturers, retailers and wholesalers) and breeders (gene banks, seed companies, etc.). The relationships between these domains are pictured in Figure 1.

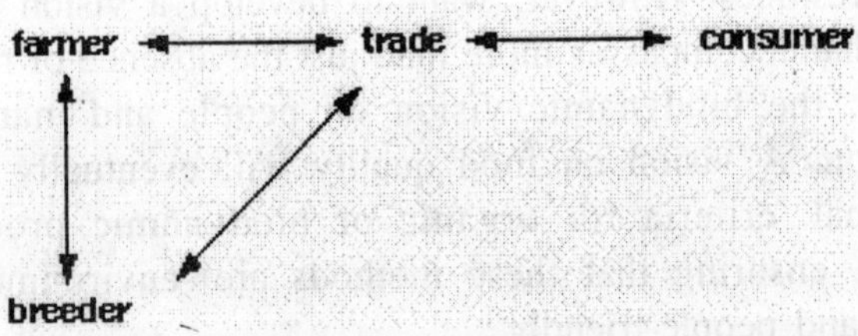

Figure 1. The three main parties in the organic plant breeding chain

Farmers

An organic farmer is dependent on organic fertilisers and nitrogen-fixing legumes for mineral inputs and this in turn makes him dependent on regional and local circumstances. From an organic point of view, the farm is a functioning organism, a natural element in a dynamic environment. Improving performance, therefore, is a matter of repeated fine-tuning and precision regulation. In this process, organic farmers gain knowledge of and experience in dealing with regional and local possibilities and limitations.

This experience explains their preference for varieties that perform well in many different growing environments and that are resilient enough to be able to adapt independently to natural seasonal variations in temperature and mineralisation. Such varieties result from breeding and selecting in specific regional and organic field situations.

Trade and Industry

Parties in trade and industry, including consumers, demand products that meet certain quality standards (product grade, keeping quality, sorting quality and size). Consumers expect organic products to be good for their health and to be produced in an environmentally-friendly manner. For that reason, they prefer fresh, preferably unprocessed, tasty produce that keeps well and is free of chemical or synthetic additives. Some consumers also value produce which has a clear link with the season, a particular region or producer. Partly in response to consumers' wishes, organic retailers are increasingly demanding varieties with species-specific and regional qualities, for example in taste, bouquet, colour and keeping quality.

Breeders

Breeding has become highly specialised, requiring specific experience with and knowledge of crossing and selection methods. Organic breeders should maintain and develop varieties in a sustainable manner, on the principle that these varieties are part of our cultural heritage. In practice, this means maintaining or if necessary increasing genetic diversity. Important conditions for the maintenance of genetic diversity are the free use and exchange of varieties among breeders, and the varieties' ability to reproduce naturally in organic growing conditions.

The most important conviction that unites the three links in the organic production chain is that products, production methods and processing methods should respect the wholeness of the living entity as much as possible, to ensure that living entities—the farm organism, the plant organism and the human organism—retain their self-regulating ability. That is why organic farmers use organic fertilisers, natural pesticides and other natural substances and methods.

Why consumers buy fresh, unprocessed produce without chemical or synthetic additives whenever they can. And why organic breeders should only work at the level of the crop and plant. DNA techniques, and to a large extent cell techniques too, occur outside the context of the whole plant in its relation with the environment. These techniques manipulate bits of plants in sealed, artificial laboratory conditions. In view of the goal of organic breeding, they are an ecological detour.

Each party in the breeding process contributes specific knowledge and wishes. Close collaboration between the parties is a must if they are to overcome possible conflicts of interests and agree on a set of breeding goals. In particular, supermarkets' demands regarding product characteristics, which in effect are demands on breeding, can conflict with what organic farmers (or consumers) want. For example, supermarkets only want to sell carrots that are 10-15 cm long, have a uniform shape and a smooth skin. Such carrots can only be produced using hybrid varieties that are lacking in taste.

Organic farmers, too, may have wishes that differ from what conventional breeders are used to. High yielding Golden Delicious is a popular cross-parent in apple breeding. But Golden Delicious is highly susceptible to disease, an undesirable characteristic from an organic point of view.

Organic apple breeding programmes, therefore, should not be based on crosses with Golden Delicious. An interactive approach to breeding may provide that intensity of collaboration which is so crucial to organic agriculture. A good start to interaction is to inform partners in the chain about various processes; farmers and trade and industry could learn about organic breeding projects at open houses or during site visits. Breeders should be able to justify their actions to downstream parties, because their products serve the rest of the chain, after all. Variety groups and selection on working organic farms could also play an important role in interactive breeding.

Criteria at Farm Level

Most principles of organic agriculture pertain to farm management. Closed production cycles, natural self-regulation and agro-biodiversity are the foundations of a stable agro-ecosystem, which forms the basis of organic production because it achieves the best balance of sustainability and economic feasibility.

Closed Production Cycles

Most farms that convert to organic production in the Netherlands are specialised farms that depend on mineral inputs (arable production) or fodder and straw inputs (livestock production). In the future, however, organic farms will have to have closed nutrient cycles to reduce losses to an absolute minimum. This can be realised by reuniting arable and livestock production, either by establishing new, mixed farms or by intensive collaboration between specialised farms according to the Partner Farm concept. Such closed farm systems raise new challenges for plant breeders. For example, in closed mixed systems fodder crops must be grown that fulfil both the energy requirement of typical mixed-system livestock and the straw requirement for slurry composting.

Self-regulation

Soil fertility plays a key role in a self-regulating agro-ecosystem.

Various aspects of soil fertility form the basis of healthy crop growth and as such are at the heart of organic farm management. An effective regime of fertiliser applications, biological nitrogen fixation and a carefully planned crop rotation scheme are requisites for good soil fertility.

The decision to grow a certain crop is based on market demand and on-farm conditions; care should be taken at all times to prevent depletion of the soil. Compared to conventional farms, organic farming conditions are more diverse, which means that there is at least as much diversity in farm management styles. These different styles of farm management all require their own varieties. This is a task for plant breeders.

The greater the diversity of crops and other organisms on a farm, the greater its natural self-regulating capacity. Natural features such as ditch bank vegetation, hedges and flowers provide shelter and food for the natural enemies of parasites. A rich varied soil life, legumes and cover crops contribute to soil fertility and prevent plagues of soil pathogens. Finally, many insects give farmers a helping hand with crop pollination.

Criteria at Plant Level

When the organic criteria at farm level, above, are extrapolated to the level of the plant, get three criteria that can be used to in organic breeding programmes: natural reproductive ability, ability to adapt independently to the environment and genetic diversity while respecting natural species authenticity and species characteristics.

Natural Reproductive Ability

One type of closed production cycle at plant level is the cycle from seed to plant to seed, which have called natural reproductive ability. Several reasons for preference for varieties that are naturally fertile in organic growing conditions, or that at least have the potential to reproduce naturally.

First, in an organic plant breeding system, all activities take place in organic growing conditions, from maintaining parent lines to crossing to selection. This automatically precludes any steps—such as fertilisation—being carried out in laboratories, so quite simply plants must be able to reproduce naturally.

Secondly, in organic agriculture a plant's reproductive ability is regarded as a sign of vigour. Vigour is a vital quality which is indicative of health and nutritive value. That is why organic farmers prefer to use varieties that can complete their life cycle in organic conditions.

Finally, natural reproductive ability ensures sustainable use of the cultivar, as progeny evolves to keep pace with the times and changing conditions.

Why do genetically identical seeds sown in different plots each grow into a unique plant? According the concept of "plant x environment interaction", the environment (available nutrients, soil structure, moisture content, air temperature, etc.) and changes to that environment during the plant's growth have just as much influence on a plant's appearance (phenotype) as its genes. In other words, a plant (genotype) is continuously interacting with its environment.

Conventional breeders recognise this to some degree as reflected by their breeding programmes, where trial plots or breeding stations are situated in different geographical regions around the world, so that plants can be selected for local climate conditions. In doing so, conventional breeders seek to develop varieties with a fairly stable phenotype. At the same time, however, conventional farming conditions are adapted as much as possible to the needs of the varieties used. Applications of chemical fertilisers and pesticides reduce inter-local differences in soil fertility and resistance to disease and thus increase a variety's area of distribution.

In organic agriculture, on the other hand, environmental diversity is respected and farmers try to optimise their use of natural resources with minimal synthetic, external inputs.

Varieties with a high adaptive capacity perform best in these conditions. In addition, organic farmers have only limited possibilities for intervening when things threaten to go wrong during the growing season. That is why varieties for organic farming must be resilient and flexible. This does not mean that adaptation to local conditions is not important in organic agriculture—it is.

The challenge for organic agriculture lies in finding the balance between varieties that are adapted to specific, local growing conditions ("specialised") and varieties that can adapt to different conditions, that have high resilience and therefore perform well in many different environments ("generalised"). What organic farmers really want are varieties adapted to local conditions (specialists) with a high resilience (generalists).

Plants have different defences against cross-pollination with other species, the ability to distinguish pollen on the stigma. These mechanisms help to preserve a species' identity. An often heard question in discussions on the use of biotechnology in plant breeding is whether the line of species authenticity may be crossed. This is a difficult question, especially as there is no clear biological definition of what a species is. Biologists generally use the term species when the following two conditions apply to a group of individual organisms:

- the individual organisms have a high degree of similarity;
- the individuals can produce fertile offspring through sexual reproduction.

There is no black-on-white definition for a species and some plant species are so flexible that they defy any attempt at definition. Take, for example, the wild cabbage, which through domestication has been cultivated to many different forms. And at the other end of the spectrum, spontaneous cross-pollination can give rise to new species, too. The mere fact that spontaneous crosses do occur naturally among some plants is no reason to freely allow such practices in plant breeding.

In nature, spontaneous crosses are the exception rather than the rule and they always occur within set, natural limitations. In organic plant breeding, it may be acceptable to cross the line of natural species authenticity provided fertilisation and embryo growth occur naturally on the plant (i.e. there is no need to resort to techniques to incorporate change at cell level) and progeny resulting from the cross are fertile.

Plant quality

A species, regardless of whether it is wild or cultivated, has distinct characteristics which set it apart from other species. These characteristics may pertain to a plant's shape, colour, keeping quality and taste. Although taste is a fairly subjective characteristic, taste does give a variety "character". The perception of taste is subjective, but the description of taste may be objective. These characteristics are partly hereditary and partly the result of environmental influences. A variety's characteristics may be changed by breeding, sometimes for the worse as in conventional breeding where too strong a focus on quantity has in many cases diminished quality.

Region-specific characteristics

If a cultivar is adapted to or used in a particular environment, it may subsequently acquire specific characteristics resulting from the interaction with that environment. Such region-specific characteristics could be used to promote regional products like bread and beer.

Nutritional value

Most cultivars are eaten by people, who believe that fruit and vegetables are good for them because of their nutritional value. However, the concept "nutritional value" is not clearly defined. The general assumption is that "healthy crops make for healthy eating". Presumably then, plants that have grown in a well-balanced, steady manner without the use of pesticides and harmful substances (natural toxins, pesticide residues) and which contain sufficient proteins, minerals and vitamins are healthy. Well-balanced or harmonious growth takes the middle road between pervasion and stagnation.

In biodynamic circles in particular, harmony in the plant's gestalt and ability to mature are considered important factors of food quality. A plant is said to have developed harmoniously if under the given conditions, its development has been optimal in each phase of growth. In wheat crops, the preference goes to varieties with a clear distinction between vegetative and generative development, which is expressed in the plant's overall development and morphology. Species-specific characteristics such as keeping quality, taste, colour and bouquet will be optimally developed if a cultivar is allowed to grow in a harmonious manner.

The sum of these quality characteristics is called "maturity". Since only ripe or mature crops are healthy to eat, ability to reach maturity could be adopted as a selection criterion in organic breeding. However, further research is required to develop the concept of nutritional value and how it may relate to plant breeding.

Socio-economic Criteria

In modern plant breeding, there is a strict division between breeders whose task is to produce new varieties and farmers whose task is to use them. A less distinct division with more consideration for farmer-breeder interaction is believed to be more productive in organic agriculture. Another limitation of conventional plant breeding is its focus on standard varieties and strains that do not take into account the great variety of organic farming conditions. These limitations are economic, institutional, technical and conceptual in nature.

Economic, because the costs of a breeding programme and the commercial objectives of propagators result in varieties that cover the largest possible geographical range. Institutional, because of the limited possibilities for breeding crops for specific environments. Technical, because techniques are used which conflict with organic agricultural principles.

And finally, conceptual because of conventional breeders' product-oriented approach common to technology development which largely ignores alternative uses of new techniques and their products. These limitations must be overcome if breeding programmes are to be developed which better meet the needs of organic farmers. To this end, breeders should seek closer ties with organic farming practice.

Farmer-breeder interaction

An organic breeding system involves selecting varieties in *situ*, that is on a working farm. The process of adapting to local production conditions is ongoing in organic breeding. Farmers can play an active part in this selection process, but this does require the relationship between farmers and breeders to be redefined. In this new interactive relationship new propagating material and new varieties could be developed. The interactive approach to breeding is quite common in developing countries, where it is known as participatory plant breeding.

Organic agriculture does not oppose specialisation and professionalism provided that there is ongoing interaction at a high level of incorporation, as for example when specialised farms form collaborations. If a farmer does not have the skills required to propagate seed and elects to outsource that part of the plant-seed cycle, it is vital that he or she keeps in touch with the propagator and vice versa. Interaction ensures an optimal use of all available knowledge and experience and ultimately leads to a better breeding result.

Selection of Breeding Techniques

One of the main principles of the vision on organic plant breeding is that there should always be interaction between plant and environment, and between farmer (and trade and industry) and breeder. Ideally, breeding programmes should be developed in consultation with partners in the organic production chain and selection should be carried out by farmers in organic growing conditions. This would result in varieties optimally adapted to organic growing conditions with characteristics that would satisfy all partners in the chain.

Techniques at plant and crop level

Preferably, interactive organic plant breeders should only use techniques at the level of the plant and crop. Parent lines would have to be maintained in organic conditions, just as progeny would be selected and propagated in organic plots. Techniques at the level of the plant or crop allow farmers to carry out the selection process themselves on their own farm or in their own region, though possibly in consultation with a professional breeder. In this way, organic management and regional influences would be incorporated in the selection process.

Hybrids

When using hybrids, F1 progeny must be maintained by repeatedly crossing the parent lines, since generative propagation of the F1 defeats the purpose of hybridisation. This end-of-the-line nature of hybrids is why some—chiefly biodynamic—growers and breeders object to their creation or use. Nevertheless, biodynamic farmers are known to use hybrids simply because seed-propagating varieties with comparable yields are not (yet) available or because buyers' demands regarding uniformity and size can only be met by using hybrids. The objection stated above is not shared to the same extent by ecological growers, although they, too, would welcome improvements in seed-propagating varieties as well as a more relaxed attitude on the part of buyers.

Seed-propagating varieties could equal hybrids in uniformity and yield if they were to become the focus of breeding programmes. Hybrid progeny cannot be generatively propagated, which thus effectively keeps farmers from collecting seed themselves. Although this may be an incentive for breeders to invest in hybrids, such varieties do not serve the long-term interests of organic farmers. For this reason, more effort should be put into the development of seed-propagating varieties. The use of hybrids in organic agriculture is justified in the following situations:

— comparable seed-propagating varieties are not available;

— homozygous lines still have enough vigour to be propagated in organic growing conditions;

— the F1 is fertile and can be used as a cross-parent, i.e. no cms without restorer lines.

Techniques at cell level

Cell techniques such as the culture of embryos and ovaries are used to incorporate characteristics of wild plants in cultivars. Embryos are cultured after pollination and seed formation on the plant because there is a reduced chance that they will grow successfully on the plant. In such cases, the embryos or ovules are isolated from the plant and placed in a culture containing all the necessary nutrients and hormones. These crosses could be created in natural circumstances, but it would be a case of beating the odds, which would require a much greater number of plants.

Techniques at cell level bypass the direct interaction between the plant and its (future) growing environment. From a biological perspective, the cell is the smallest living entity and hence cell techniques could be used without violating the organic principle to work only with living entities. On the other hand, the smallest unit of life encountered by farmers in daily farming practice is the whole plant in relation to its environment. From this perspective, cell techniques are an ecological detour and as such their use in organic breeding continues to be a point of controversy.

In conventional breeding, however, cell techniques have been used since the 1930s. For some crops, such as tomatoes, they have become a crucial step in the maintenance of the cultivar. In effect, cell techniques are synonymous with breeding programmes based on specific, hereditary, usually monogenetic resistance or characteristics.

In organic breeding, however, we are not interested in incorporating absolute resistance, but in developing varieties that are better able to cope with existing growing conditions. Such a plant might have deeper or more elaborate rootage to survive plagues of insects or fungi. Organic breeding objectives are based on a different concept of plant health. More research is required to work out the operational details of this concept and to enable the development of a breeding system that is less dependent on cell techniques.

An immediate ban on cell techniques that have been used for decades might do more harm than good for some crops and varieties, and have dramatic economic consequences for farmers. Obviously, therefore, satisfactory alternatives will first have to be developed. Establishing a transitional period of at least ten years during which varieties with characteristics resulting from cell techniques may be used in organic farming and breeding while at the same time alternatives are developed that are more in keeping with organic plant health principles. A condition for using such varieties in organic agriculture is that they have the potential to reproduce naturally in the organic growing environment.

In addition, characteristics which have been incorporated using cell techniques should also be receptive to traditional cross methods. In order to stimulate the development of organic breeding programmes which are not dependent on cell techniques, propose setting up a special "green mark" programme in which participating breeders pledge to not use cell techniques.

Varieties or lines produced with cell techniques and originating from before a certain date may be used as foundation stock. Varieties and lines produced with cell techniques after 1-1-2001 would not be allowed in green mark breeding programmes.

Protoplast fusion

Protoplast fusion goes one step further than the techniques described above. Protoplast fusion is carried out with loose, unconnected cells that are in no way recognisable as plant material. This technique is applied when two species differ so much that a successful cross could never be achieved by natural methods. This is also a major reason for rejecting the technique in organic plant breeding. The consequences of rejecting this technique are not that far-reaching, as it is fairly new in conventional breeding, too. Protoplast fusion is primarily used to transfer cytoplasmic male sterility from one species to another (for example, from radish to cabbage). The technique has been patented, which is another reason to not use it in organic breeding programmes.

Cytoplasmic male sterility (cms)

Organic agriculture rejects the use of techniques that negatively affect a plant's natural reproductive ability. The sector particularly opposes cms hybrids without restorer genes. There are plenty of alternative seed-propagating varieties or hybrids without cms. Unfortunately, the use of cytoplasmic male sterility in modern hybrids is growing.

Although cytoplasmic male sterility occurs naturally in some species, breeders are currently using protoplast fusion to transfer the characteristic to other species as well. These new hybrids are subsequently patented. The absolute cms in cabbage, for example, originates from radish. Cms hybrids differ from non-cms hybrids in a number of ways:

— the advantage of using cms-hybrids in practice is that inbred plants (deviants) do not occur in the field as is the case with non-cms hybrids (sometimes more than 10%);

— other breeders cannot acquire the mother line by searching for inbred plants because these do not exist; this gives the breeder who owns the mother line a competitive advantage;

— other breeders cannot incorporate any of a cms hybrid's hereditary characteristics in their own breeding material; the breeder of the cms hybrid has exclusive access to the gene pool, so that he is the only one who can incorporate the selected characteristics in marketable varieties and profit from them. Ultimately, however, cultural inbreeding will occur and this may lead to genetic erosion. These practices therefore go against the grain of organic agriculture.

Why can cms plants (without restorer genes) never be used as cross parents? Imagine a breeder who sees a promising cms hybrid and wants to cross it with his own lines or plants. The cms plant (plant A) is male sterile (does not produce functioning pollen) and can therefore only be used as a mother plant. Crossing plant A with a fertile father (plant B) yields plants that have a 50% share of genes from each parent, including the cytoplasmic male sterility as that is always inherited from the mother. The only thing the breeder can do now is cross the progeny with a fertile plant. This F_2 generation will have inherited 25% of plant A's genes, again including the cms characteristic. Eventually, the breeder will get a generation which has 0% of plant A's genes, but these plants will still be male sterile.

The only characteristic that can be transferred from a cms cross-parent is the cms characteristic, and if that is what you want, one cross is all it takes. Every other characteristic is lost in the backcross. And because plant A is sterile, there is no way of using it other than as the mother plant. The free exchange of varieties, hybrid or otherwise, will always form the basis of sustainable breeding. This was one of the primary considerations that led to the establishment of breeders' rights for plants, as an alternative to patenting. If every single variety was patented, breeding would be made impossible. But cytoplasmic male sterility, too, makes breeding impossible.

Breeders who work with cms can use other breeders' non-cms varieties, but they in turn cannot use the cms varieties. There is no mutual exchange of benefits or sharing of profits. This one-way breeding is a form of parasitism. Breeders have a responsibility to maintain cultivars because they are part of our cultural heritage, resulting from centuries of cultivation and breeding. These cultivars cannot be maintained without natural reproduction and genetic transfer.

As cms becomes more and more common in modern varieties, fewer varieties will remain for use in breeding programmes. Eventually, modern varieties will no longer be able to be crossed to broaden the genetic basis of a breeding programme. Programmes will stagnate due to the erosion of genetic diversity. Thus, although cms may seem like an attractive way of getting ahead of the competition, it is ultimately a dead-end road. In effect, each cms hybrid is like a dead twig on a cultivar's family tree.

Breeders will need restorer genes for the infusion of new genetic material. These genes could be transferred from one species to another through genetic engineering. But it is likely that this quality will be patented just as cms was and therefore only be available for cms breeders.

Breeders who do not use cms will always be in an underdog position and on top of that, will end up in a cut-throat competition with each other. If there are no restorer genes, these breeders will not be able to incorporate the genes in cms varieties.

At the same time, however, cms breeders will gratefully cash in on each improvement in non-cms varieties. Breeders who use cms techniques are not taking their responsibility to maintain cultivars and in the long term they will drive non-cms breeders into economic extinction. The question is whether, in the current system of breeding, non-cms programmes can feasibly co-exist with cms programmes.

Artificial induction

The artificial induction of mutations in DNA by radiation, chemical mutagens or other means is completely random and largely harmful. In organic plant breeding, direct manipulation at DNA level is undesirable. Artificial induction is therefore unsuitable for use in organic breeding. An immediate ban on this technique will have little impact, as the technique is currently rarely used in conventional breeding. The same applies to radiated mentor pollen.

Techniques at DNA level

Genetic modification is not allowed in organic agriculture. This is justified on the basis of the following organic principles:

— It is a one-dimensional, drastic intervention in a plant's genetic make-up which upsets the plant's relation to the environment. The desired DNA can only be incorporated by first reducing the plant to cell level and then "reconstructing" it using tissue culture techniques. This destroys the historical plant x environment relationship.

— It is a reductionist method that increases the chances of undesirable and unpredictable ecological and health risks in the long term.

— It is a capital-intensive type of breeding in which multinationals hold patents on living organisms and impede a free exchange of genetic material, thus reducing genetic and cultural diversity.

DNA diagnostic techniques

Genetic engineering has also enabled selection at DNA level without actually modifying plants' genetic make-up. These techniques are based on biochemical and molecular markers and may complement field selection methods. The potential of these indirect selection techniques should not be exaggerated, however. Although there is no reason why these techniques should not be applied in organic breeding, selection in the field will always be the primary method of selection because it incorporates the plant x environment interaction.

11

APPLICATION OF RICE GENOMICS TO PLANT BREEDING

As the main source of nutrition, 50% of the human population depends on rice (*Oryza sativa* L.). Rice has a small genome (430 Mb) relative to other Gramineae plants, and its genome size is about three times larger than that of *Arabidopsis*, a model plant of dicotyledons (dicots). Last decade, technological innovations in science enabled dramatic advancements in the field of plant genomics (genome science). Many rice genome projects have been launched, and they are providing very useful information for plant biology and plant breeding.

Rice genomics is a significant area of research with applications for plant biology and breeding, and its possible role in overcoming problems associated with climate change and food production should be resolved. The rice genome shows apparent syntenies with many other grasses, such as wheat, barley, and maize. These syntenies suggest that rice genomics has implications not only for rice breeding but also for other cereal crops.

Over the past 100 years, rice breeders and geneticists have accumulated a large number of rice mutants and a rice classical genetic map using phenotypic markers of these mutants has been developed. The compact nature of the rice genome provides a distinct advantage in gene isolation and genomic sequencing as opposed to other cereal crops, and the results of rice genomics can be directly applied to cereal breeding due to their syntenies, as described above. For these reasons, rice has been selected as a target species for genome research by a number of groups. Developments in DNA technology have enabled the detection of DNA diversity in plants.

In 1988, a research group at Cornell University first applied the Restriction Fragment Length Polymorphisms (RFLPs) technique to construct a rice linkage map, and the group subsequently developed a fine linkage map with 726 DNA markers. A Japanese group has also constructed a rice RFLP linkage map. To improve understanding of the rice genome, the Japanese government funded a seven-year Rice Genome Research Programme (RGP) between 1991 and 1997. The RGP was initiated as a collaboration between the National Institute of Agrobiological Resources (NIAR) and the Society for Techno-innovation of Agriculture, Forestry and Fisheries (STAFF) under the Japanese Ministry of Agriculture, Forestry and Fisheries (MAFF). This programme was composed of four main projects: cDNA analysis, genetic mapping, physical mapping, and bio-informatics.

In the programme, a large number of cDNAs were isolated, sequenced and used as DNA markers for constructing a linkage map and a physical map of Yeast Artificial Chromosomes (YACs). Then, the rice classical genetic map with morphology markers and the RFLP linkage maps with molecular markers were integrated. The accumulation of DNA markers, together with the construction of the linkage map and physical map, made it possible to isolate important genes. The *Xa21* gene, which confers resistance against the bacterial pathogen *Xanthomonas oryzae* pv., was first isolated using a positional cloning strategy. Subsequently, several genes for important agronomic traits have been cloned by the same strategy.

Rice, having one of the smallest genomes among the food crops, is the focus of basic crop research in many countries and will be the first food crop to be completely sequenced. Progress in understanding rice molecular genetics, established in the first phase of the RGP, and technological innovations in genomics make a reality of the rice genome sequencing projects. Analysis of the human genome promises to help elucidate the characteristics of inherited disease and provide the possibility of overcoming these diseases. It is equally important that staple foods necessary to sustain human life be chosen as targets for genome analysis.

The International Rice Genome Sequencing Project (IRGSP) was launched in 1998 with ten countries participating. August 2001, thirteen research groups in eight countries are involved in the project and share the chromosomes for sequencing: Japan, Rice Genome Research Programme (RGP), chromosome 1, 2, 6, 7 and 8; Korea, Korea Rice Genome Research Programme (KRGRP), chromosome 1; UK John Innes Center, chromosome 2; USA, Clemson Univ. (CUGI), Cold Spring Harbor Laboratory (CSHL), Washington University School of Medicine-Genome Sequencing Center (GSC), The Institute for Genomic Research (TIGR), Plant Genome Initiative at Rutgers (PGIR), Genome Center of Wisconsin, chromosome 3, 10 and 11; China, National Center for Gene Research Chinese Academy of Sciences (NCGR), chromosome 4; Taiwan, Academia Sinica Plant Genome Center (ASPGC), chromosome 5; Thailand, National Center for Genetic Engineering and Biotechnology (BIOTEC), chromosome 9; and France, Genoscope, chromosome 12.

The IRGSP immediately releases the rice genomic sequencing data in public databases such as the DNA Data Bank of Japan (DDBJ), GenBank and the European Molecular Biology Laboratory (EMBL). Progress on sequencing is also exhibited in the RGP database. The sequencing data, rice linkage map, and physical map are integrated in the RGP database, the Integrated Rice Genome Explorer. In 2000, a private company, Monsanto, made 60% of its rice genome sequences public. They also released about 7,000 sequence data including microsatellites. These data will accelerate the completion of the whole rice genomic sequencing project and make it easier to design DNA markers for specific positions on the rice chromosomes.

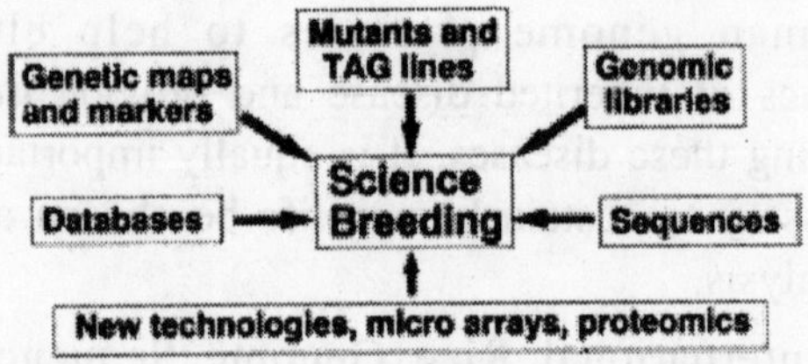

Figure 1. Application of rice genomics to science and breeding.

In 2001, another private company, Syngenta, announced that it had finished sequencing of the rice genome, but the data have not yet been placed in the public domain. The data provided by rice genomics give very powerful information for rice biology and breeding (Figure 1).

Application of Rice Genomics

To date, a great number of rice mutants have been reported, and they have also been produced artificially by using mutagens such as irradiation, chemical compounds and DNA insertion, including T-DNA, *Ac/Ds* transposon, and retrotransposon, *Tos17*. These mutants are very significant materials for functional genomics, enabling to study and predict the function of genes. Due to progress in the development of genomic tools, molecular markers and genomic libraries, and the accumulation many kinds of sequence data (Figure 1), gene cloning using various mutants has become more and more realistic.

The application of rice genomics to molecular biology, using the analysis of the gibberellin (GAs) biosynthetic and signal transduction pathways in rice as an example. GAs were first identified in the process of studying rice. A fungal pathogen of rice, *Gibberella fujikroi*, causes an abnormal stem elongation known as Bakanae-disease. Kurosawa discovered that this abnormal stem elongation is due to a water-soluble compound, gibberellin, produced by the fungus. GAs are a large family of tetracyclic diterpenoid plant growth regulators reportedly associated with a number of plant growth and development processes such as seed germination, stem elongation, flowering, fruit development, and regulation of gene expression in the cereal aleurone layer.

GA-related mutants in plants show dwarf or elongated phenotypes, and these mutants are crucial for elucidating the regulatory mechanisms governing the GA biosynthetic and signal transduction pathways. Because dwarf characteristics are favored in plant breeding, the study of these characteristics has applications not only for understanding basic plant biology but also for molecular breeding. Many GA-related mutants have been isolated from numerous plant species and can be roughly classified into two categories: GA-sensitive and GA-insensitive. A GA-sensitive mutant responds to exogenous GA because it cannot produce GA, or it produces insufficient GA due to a deficiency in genes encoding GA catalytic enzymes.

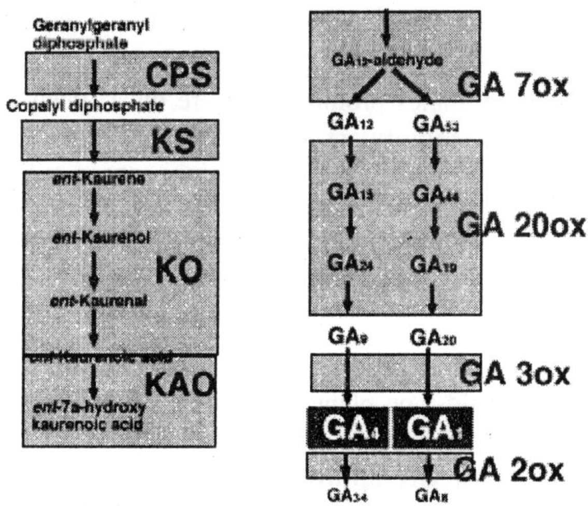

Figure 2. Gibberellin biosynthetic pathway in higher plants.

On the other hand, a GA-insensitive mutant does not respond to exogenous GA, and genes related to GA-insensitivity may be associated with GA signal transduction. The GA biosynthetic pathway has been well studied in plants, and at least seven enzymes are known to catalyse GA synthesis in plants (Figure 2).

GA Biosynthetic Pathway

In rice, 61 dwarf mutants have been registered to date. However, only two of these mutants, *d18* and *d35*, are characterised as having defects in their GA biosynthetic pathway. Bio-active GA_1 and GA_4 are catalysed from geranylgeranyl diphosphate (GGDP) by sequential action of cyclases in the plastids, membrane-associated mono-oxygenases in the endoplasmic reticulum, and soluble 2-oxoglutatate-dependent dioxygenases (2ODD) in the cytosol (Figure 2).

The isolation of the 3β-hydroxylase (GA3ox) gene, whose product catalyses the final step in GA biosynthetic pathway, from GA_{20} to GA_1 and GA_9 to GA_4. *d18* has been characterised as a GA-sensitive dwarf mutant, and four independent alleles have been identified, of which two are strong alleles and *d18*-AD, and two are weak alleles and *d18k*.

The strong alleles promote a severe dwarf phenotype, the weak alleles a semidwarf phenotype. Physiological and biochemical experiments have been mainly carried out using the *d18-dy* mutant. The dwarf phenotype of the *d18-dy* mutant is rescued by the application of GA_1, but not by GA_{20}. Seedlings of this mutant are deficient in GA_1 and accumulate its immediate precursor, GA_{20}.

The metabolism of GA_{20} to GA_1 is much lower in *d18-dy* plants than in wild-type plants. Based on these observations, it has been predicted that the *d18-dy* gene encodes a GA 3β-hydroxylase. Two GA 3β-hydroxylase genes, *OsGA3ox1* and *OsGA3ox2*, from rice by screening a genomic library with a DNA fragment encoding the conserved amino acid sequence of GA 3β-hydroxylase. Both proteins produced by the cDNAs in *E. coli* show 3β-hydroxylase activity for the steps GA_{20} to GA_1 and GA_9 to GA_4.

Molecular and linkage analyses have mapped the *OsGA3ox1* gene to the distal end of the short arm of chromosome 5, and the *OsGA3ox2* gene to the distal end of the short arm of chromosome 1, which corresponds to the *D18* locus.

The association of the *OsGA3ox2* gene with the *d18* locus has been confirmed by sequence and complementation analysis of the *d18* allele. Complementation of the *d18-AD* allele with the *OxGA3ox2* gene results in transgenic plants with a normal phenotype. Northern hybridization has revealed that the genes are expressed in an organ specific manner, with the highest level of *OsGA3ox1* expression occurring in unopened flowers and the highest expression level of *OsGA3ox2* in elongating leaves.

Therefore, the two 3β-hydroxylases may be involved in distinctive growth and development events in rice plants. Recently cloned six rice genes catalysing all steps of the GA biosynthetic pathway except *GA7ox*, and also identified six dwarf mutants which are functionally disrupted in each of the six genes. These results have implications for molecular breeding.

GA Signal Transduction Pathway

The mechanisms of signal transduction triggered by GAs are still poorly understood compared to the GA biosynthetic pathway. To elucidate the mechanism of GA signal transduction, the rice GA-insensitive dwarf and slender mutants have been screened. So far, three genes the *d1*, *slr1* and *gid1* genes have been cloned using positional cloning, undoubtedly a very strategic way to isolate a gene. Three good examples of the functional analysis of GA signal transduction using rice genomics.

d1 mutant: The rice dwarf mutant Daikoku, carrying the *d1* gene, was first isolated as a spontaneous mutant that was not only short, but also had broad, dark green leaves, compact panicles, and short-round grains. These phenotypes are all induced by a recessive allele (*d1*) of the *D1* gene and are thought to reflect aberrant physiological and biochemical pathways in plant growth and development. The rice d1 mutant was classified as gibberellin-insensitive.

The *D1* gene, cloned by positional cloning, encodes the α subunit of a *G*-protein. It is well known that G-proteins play an important role in signal transduction in animals and microbes. The fact that the rice mutant *d1* is GA-insensitive suggests the *D1* gene may be associated with GA signal transduction. The putative function of the α subunit of the G protein (Gα) and the predicted function of the *d1* gene are very consistent.

GA signaling in *d1* and the role of the Gα protein in the GA signaling pathway. Compared to the wild-type plants, α-amylase activity in the aleurone cells of *d1* is greatly reduced. The GA_3-treated aleurone layer of *d1* has a lower expression of *Ramy1A*, which encodes α-amylase, and of *OsGAMYB,* which encodes a GA-inducible transcriptional factor, and no increase in the expression of Ca^{2+}-ATPase relative to the wild-type. However, in the presence of high GA concentrations, α-amylase induction occurs even in *d1*.

The GA sensitivity of second leaf sheath elongation in *d1* is similar to that of the wild-type plants in terms of dose-responsiveness, but the response of internode elongation to GA is much lower in *d1*. Furthermore, *OsGA20ox* expression is up-regulated, and the GA concentration is elevated in the stunted internodes of *d1*. Together, these results suggest that *d1* affects a part of the GA signaling pathway; namely, the induction of α-amylase in the aleurone layer and internode elongation.

In addition, a double mutant between *d1* and *slr1* revealed that *SLR* is epistatic to *D1*, supporting the notion that the Gα protein is involved in GA signaling. However, the data also provide evidence for the presence of an alternative GA signaling pathway not involving the Gα protein. It is proposed that GA signaling via the Gα protein may be more sensitive than that of the alternative pathway, as indicated by the low GA-responsiveness of this Gα independent pathway.

slender (slr) Mutant

The rice *slr* mutant shows a slender phenotype with an elongated stem, leaf sheath, and blade, similar to that of rice plants that have been exogenously treated with GA_3. Consequently, the slender phenotype seems to be the constitutive GA response phenotype, as if it is saturated with GAs. In the *slr* mutant, elongation is unaffected by treatment with Uniconazole, a GA biosynthesis inhibitor. Also, GA-inducible α-amylase is produced in the aleurone cells without GA application, and the endogenous GA content is lower than that of the wild-type. These results indicate that the *SLR1* protein is associated with GA signal transduction as a negative regulator.

The *SLR1* gene and found that it encodes a putative repressor protein for the GA signaling pathway homologous to Arabidopsis GA-Insensitive (*Gai*) / Repressor of *gal-3* (*Rga*), wheat Reduced height (*Rht*), and maize *dwarf-8* (*D8*). Transgenic rice plants that overproduce *SLR1* show the dwarf phenotype, supporting the idea that *SLR1* functions as a negative regulator of GA signaling.

Dominant alleles in the Arabidopsis *GAI,* wheat *Rht-B1/ Rht-D1* and maize *D8* loci confer GA-insensitive mutants with the dwarf phenotype. Molecular cloning of Arabidopsis *GAI* has demonstrated that the in-frame dele-tion of its *N*-terminal domain occurs in the gai mutant. Similarly, wheat *Rht-B1/Rht-D1* and maize *D8* have mutations in their N-terminal domains as in *GAI*.

According to the dominant phenotype caused by the gai mutant protein, it has been suggested that the native *GAI* product represses the action of GA and that its repression can be released by GA. Peng et al. also predicted that the internal deletion of the *GAI* protein in the gai mutant is resistant to the GA signal. If this were the case, loss-of-function mutants of the *GAI* product should show a constitutive GA response with the slender phenotype regardless of the presence or absence of GA. However, plants with loss-of-function alleles of gai only show a slight reduction in GA dependency.

The absence of a clear phenotype in the *GAI*-knockout plants is thought to be due to the presence of redundant genes. Indeed, the *RGA* gene has a highly similar structure to that of *GAI*, and its loss-of-function does not show a typical constitutive GA response phenotype but partially suppresses the dwarf phenotype conferred by the GA-deficiency mutation, *ga1-3*.

In contrast to these cases in Arabidopsis, the rice genome has only one gene encoding an orthologous protein to *GAI/RGA/Rht/d8*, and consequently rice plants with a loss-of-function allele of *SLR1* show a constitutive GA-responsive phenotype. Such non-redundancy in rice should provide an advantage for studying the GA signal transduction pathway. Using transgenic plants that overproduce a fusion protein consisting of *SLR1* and green fluorescent protein (SLR1-GFP), *SLR1* functions in the nucleus to repress GA action and that a GA signal causes *SLR1* protein levels in the nucleus to decrease, resulting in the induction of stem elongation.

Recently, it has also been demonstrated that *Arabidopsis RGA* and *GAI* proteins occur in the nucleus and that the proteins disappear upon the application of GA_3, as does the rice *SLR* protein.

Isolation of New Genes

Several rice GA-insensitive dwarf mutants from mutagenized M_2 lines in order to elucidate the mechanisms underlying the GA signal transduction pathway. Recently, selected many mutants showing severe dwarfism, dark green leaves and sterility, which are similar to the severe allele of the rice GA deficient mutant *d18*. The GA sensitivity of these dwarf mutants against three indicators of GA responsiveness, that is, elongation of the second leaf sheath, α-amylase induction in aleurone cells, and expression of *GAC20* oxidase.

Seven of these mutants showed no second sheath elongation or α-amylase induction in response to GA treatment. Highlevel expression of the *GAC20ox,* which is negatively regulated by active GA in a feedback manner, was observed in the mutants with or without GA treatment. All of these observations indicate the GA-insensitivity of these dwarf mutants.

Gibberellin Insensitive Dwarf 1(gid1) Mutant

As a first step towards gaining a better understanding of GA signaling in plants, the gibberellin insensitive dwarf 1 (*gid1*) mutant for detailed characterisation. There are five *gid1* alleles. In *gid1-1* to *gid1-5*, no elongation of the second leaf sheaths and no α-amylase induction in aleurone cells, can be observed at any level of GA concentration. The *GAC20ox* is highly expressed in *gid1-1*, and the accumulated level of GA_1 is 100 times higher than that found in wild-type plants.

All of these findings demonstrate that *GID1* encodes a positive regulator of GA signal transduction. To elucidate the molecular function of *GID1*, have cloned *GID1* by positional cloning and identified that *GID1* encodes a novel gene. Currently investigating the relationship between the *GID1* and *SLR* proteins in GA signaling.

Application to Breeding

In the 1960s, an impending global food crisis was a major concern because it was predicted that demand for food would exceed production. The world population was expanding rapidly due to a significant decline in mortality rates resulting from advancements in modern medicine and human health care, while the availability of land for cultivation had dramatically declined as the result of desertification caused by reckless deforestation and construction.

Recognising the problem, in the 1960s the International Rice Research Institute (IRRI) developed a semi-dwarf variety, *IR8*, known as "miracle rice." Widespread adoption of *IR8* led to major increases in rice production, and a large famine was averted. The technical revolution and improvements in irrigation and fertilizers also facilitated increases in rice production. All of these advances in the 1960s constitute the so-called rice "green revolution." As the result, from 1960 to 1990 the global production of grains more than doubled. Since 1990, growth in the grain harvest has slowed dramatically. However, the population of the world is continuing to increase by 80 million people a year, with 90% of this increase occurring in the developing countries of Asia, Africa, and Latin America. Providing for population growth now requires an expansion in world grain production by 26 million tons per year. In total, world food grain production must increase by 50%.

To feed a world population of 8 billion by 2025, to act now and develop both political and scientific strategies for meeting the challenge. In scientific terms, there are various strategies for increasing food grain production including: 1) genetic improvement of crop cultivars; 2) technological improvements in irrigation and facilities; and 3) sustainable management of crops and pests, and water, nutrient, and soil resources.

In political terms, consistent public policy and fundamental support for research and basic infrastructure are required. Here, focus on the scientific strategies to improve food production. Several biotechnological approaches for increasing crop yield potential are being investigated. These include the introduction of cloned novel genes through transformation and the use of molecular marker technology. The two strategies for breeding: molecular breeding and Marker Assisted Breeding.

Genetic engineering and biotechnology hold great potential for plant molecular biology and plant breeding as they promise to expedite the time taken to produce crop varieties with desirable characteristics. Because rice breeding by conventional crossing and selections may not catch up with increasing global food demand, other more efficient strategies are required. One such strategy is the molecular breeding approach. The application of molecular studies of plant hormones to molecular breeding.

A semi-dwarf variety IR8, produced by the IRRI, increased rice production and brought about a "green revolution" in rice in the 1960s. The semi-dwarf 1 (*sd1*) gene contributed to this revolution. Around the same time, a wheat semi-dwarf variety produced at CIMMYT (Wheat and Maize Improvement Center) also enabled a green revolution in wheat. In this case, a semi-dwarf gene, Reduced height (Rht), was important for improving yields in wheat.

The Rht gene encodes an *Arabidopsis GAI* ortholog, a negative regulator of GA signal transduction. Without these semi-dwarf genes it is unlikely that the green revolutions for rice and wheat would have been possible. In general, nitrogen fertilizer increases grain production, but it also increases culm elongation and the overall height of the crop plant. Wind and rain can readily dislodge tall plants enhanced by fertilizer, causing considerable damage to the crop and yield losses.

The semidwarf rice variety IR8 responds to fertilizer inputs and produces an increased yield without culm elongation. Particularly in the monsoonal regions of Asia, where typhoons frequently occur during the yielding season, the semi-dwarf characteristic is a very important agronomic trait for crop breeding. Change the height of rice plants by controlling the expression of two GA-related genes, GA 2-oxidase and GA 3β-hydroxylase gene. To do this, employed two strategies: overexpression of exogenous GA 2-oxidase in transgenic plants and reducing GA 3β-hydroxylase gene expression (Figure 2).

In many plant species, bioactive GAs are regulated by metabolism and catabolism of active GAs. The catabolism steps, from active GA_1 to inactive GA_8 and active GA_4 to inactive GA_{34}, are catalysed by one of the 2-oxoglutarate-dependent bioxygenases (2ODD), GA 2-oxidase. A rice GA 2-oxidase gene, *OsGA2ox1*, has been cloned. Because GA 2-oxidase plays an important role in the regulation of plant growth via the reduction of endogenous levels of bioactive GAs, plant height could be controlled by modifying the GA 2-oxidase gene. *OsGA2ox1* has been introduced into wildtype plants to reduce levels of active GAs, and overexpression of the *OsGA2ox1* cDNA in transgenic rice plants inhibits stem elongation.

In the GA biosynthetic pathway, GA 3β-hydroxylase catalyses the reactions from inactive GA_9 to bioactive GA_4 and inactive GA_{20} to bioactive GA_1. The reduction of GA 3β-hydroxylase expression in an antisense fashion causes a decrease in production of bioactive GAs. The rice 3β-hydroxylase gene, *OsGA3ox2*, was isolated by Itoh.

Transgenic plants carrying an antisense construct of 3β-hydroxylase have a semidwarf phenotype (Figure 3), and have successfully produced dominant semi-dwarf rice plants. Because GAs commonly occur in plants and are used in the control of plant growth and development, these data suggest that controlling the height of any crops may be possible via the modification of GA genes.

Figure 3. Molecular breeding. 1, Producing a semi-dwarf rice plant by the controlling a GA biosynthetic gene.

Rice plant architecture

An erect leaf is a desirable agricultural trait for high yields, because plants with erect leaves can be densely planted in the field and can also perceive sunlight more efficiently under dense planting conditions. Recently isolated a novel rice dwarf mutant, *d61*, which shows the dwarf phenotype, erect leaves, and a strange elongation pattern of the internodes.

The erect leaves and dwarfism of the mutant are caused by its insensitivity to brassinosteroids (BRs). BRs are a group of plant hormones found at low levels in pollen, seeds, and young vegetative tissues throughout the plant kingdom. BRs share structural similarity with animal steroid hormones, and have been shown to regulate gene expression and stimulate cell division and differentiation. *D61* encodes a receptor-like kinase protein, including a putative signal peptide, two conservatively spaced cysteine pairs, a leucine-rich repeat (LRR) domain, a predicted transmembrane domain, and a kinase domain (KD).

Because the structure of the sub-domains of the *OsBRI1* gene is the same as the *Arabidopsis* BRs receptor, *Bri1*, and also because of its insensitivity to BRs, *OsBRI1* is considered an ortholog of *Bri1*. The *OsBRI1* protein may be associated with BRs perception, and the function of *OsBRI1* may well be important in the BRs signaling pathway.

The function of the *Bri1* protein has been studied, and it has been shown that the extracellular domain LRR of *BRI1* is essential for the perception of BRs. The intercellular domain is thought to be a KD, and KD domain of *BRI1* has a capacity for auto-phosphorylation. Recently, Wang et al. demonstrated that *BRI1* is a receptor kinase that transduces steroid signals across the plasma membrane. The *d61* mutant shows erect leaves of agronomic usefulness, but it also has severe dwarfism.

Produce erect leaf plants without severe dwarfism by manipulating the expression of OsBRI1. At first, employed an antisense strategy to inhibit the expression of endogenous OsBRI1. Transgenic plants with an antisense strand of OsBRI1 cDNA under control of the rice *Actin1* promoter showed erect leaves and a severe dwarf phenotype. Thought that the antisense RNA strand had severely reduced the function of the OsBRI1 protein, and that another strategy was needed to mildly inhibit the OsBRI1 protein and produce plants with erect leaves but without severe dwarfism. Thus, employed a dominant negative technique to slightly reduce BR signaling. Because thought that the recombinant KD might function as a dominant negative factor of endogenous OsBRI1 and restrain the essential role of OsBRI1, transformed the KD of the OsBRI1 gene into TC65 and overexpressed it under control of the rice Actin1 promoter.

The transgenic plants showed erect leaves without dwarfism (Figure 4). The phenotype of the dominant negative plants in order to determine their usefulness, obtained three lines of T3 selfed-progenies of the OsBRI1 dominant negative plant. To make an accurate quantitative comparison between the transgenic plants and wild-type TC65, the lamina joint angles of the plants were measured. Those of the transgenic plants were 0-5 degrees, and those of TC65 were 15-20 degrees.

Figure 4. Molecular breeding

Measured and compared several other characteristics, including plant height, culm length, panicle length, internode elongation pattern, tiller number, grain number, weight of 100 seeds, and fertility. However, no differences between the transgenic and wild-type plants were observed. The dominant negative plants show erect leaves, but that no other characteristics are affected, including yield.

The transgenic plants with erect leaves yield as well as the wild-type TC65, but they do not require as much space as the wild variety for planting. Computer simulations suggest that, compared to TC65, 1.5 times the number of transgenic plants can be planted in equal sized fields. These figures, however, are based on plants grown under isolated greenhouse conditions. Because many environmental conditions affect rice production, field trials of the transgenic plants are required to confirm these data.

Today, the cloning of genes and production of transgenic plants are common technologies utilised in plant science. Such technologies are very powerful and efficient strategies for producing "ideal" crop plants. Actually, some transgenic crop plants, commonly referred to as Genetically Modified Organisms (GMOs), with traits such as insect resistance and herbicide tolerance have already been commercialised after passing stringent field trials. Recently, however, concerns about the impact of GMOs on the environment and their safety to humans have received greater attention.

Scientists need to be conscious of these concerns as they continue to develop and apply technologies for efficient molecular breeding using gene modification. They will need to be able to prove the safety of GMOs for humans and the environment and consult widely with the general public on matters of concern.

Consumers can then make an informed decision about whether to accept or avoid the use of genetically modified products. In the near future the use of GMOs in everyday life will be commonplace and that this technology can help to resolve the up-coming food crisis.

In the 1800s, Gregor J. Mendel discovered the genetic basis of plant breeding through his experiments using garden peas, and he established Mendel's law. However, for centuries farmers have improved crops and vegetables by crossing and selection based on desired traits such as environmental adaptability, quality, and yield. They planted seeds from individuals bearing the preferred traits and continued further selection to fix those characteristics.

A number of plant varieties have been established in this way, and the method used by farmers was based on Mendelian principles. This conventional breeding strategy remains unchanged today, but the use of molecular markers has made it more efficient. Develop rice plants with a range of improved characteristics such as increased yield, resistance to disease, and tolerance to stresses such as drought and salinisation. To achieve these plant breeding goals, the identification and isolation of the genetic information held in rice genomic DNA is essential. The efficient Marker Assisted Breeding strategies, including Quantitative Traits Loci (QTL) approaches.

The use of molecular markers in plant breeding has several advantages over the traditional phenotypic markers previously available to plant breeders. Marker Assisted Selection (MAS) is a very convenient and efficient breeding strategy in which conventional breeding selection is carried out on molecular markers linked to target traits rather than on the traits themselves. Selection is followed by the marker genotype tightly linked to target traits. RFLP markers can be used for MAS, but they cannot efficiently screen genotypes of many plants in a short time because they require intricate and time-consuming experiments.

Many kinds of PCR-based markers have replaced RFLP markers. These include RAPDs (Random Amplified Polymorphic DNAs), Sequence Tag Site, (STS), (Simple Sequence Repeats, including Minisatellite repeats and Microsatellite repeats) SSR, (Cleaved Amplified Polymorphic Sequences) CAPS, (Amplified Fragment Length Polymorohism) AFLP, and (Single Nucleotide Polymorphisms) SNP. Handling of these PCR-based markers is very simple and efficient, making it possible to hasten the transfer of desirable genes among varieties and to introgress novel genes from related wild species. SSR markers in particular are very convenient for genotyping, mapping, cloning genes, QTL analysis and MAS. Increased recognition of MAS and diffusion of molecular markers to breeders are required for more efficient breeding.

In contrast to monogenic characteristics, such as disease and insect resistance, many important agronomic traits including yield, heading date, culm length, grain quality and stress tolerance show continuous phenotypic variation. Several genes control these traits. These polygenic characteristics, including QTLs, were previously very difficult to analyse using traditional plant breeding methods, but recent progress in the identification of molecular markers has made it possible to detect the QTL easily. Primary-mapping populations, such as F_2, Recombinant Inbred Lines (RILs), Backcross Inbred Lines (BILs) and Doubled Haploid Lines (DHLs), are often used in QTL analysis to detect the chromosomal regions controlling target traits. RILs can be developed from F_2 plants by the Single-Seed-Descent (SSD) method to form a population that is useful for the detection of QTLs. BILs can be developed from backcross populations, such as BC_1 or BC_2 using SSD.

These populations are very useful for QTL analysis, but SSD can take some time to produce populations. DHLs can be developed via the anther culture of the F_1 hybrid, and this approach is relatively fast because the genotype is fixed by the doubled haploid. These three lines are very useful plant materials for detecting important agronomic traits, but they do have limitations. In QTL analysis, even using a large number of plants and genetic markers and well-developed statistical methods, it is difficult to determine the precise location of individual QTL as a single Mendelian factor in the primary analysis. Consequently, other plant materials are required to allow to proceed with further analysis, such as fine mapping and characterisation of target QTLs. To achieve high resolution mapping of a QTL, construction of Nearly Isogenic Lines (NILs) harboring the QTL is required.

The NILs are constructed by backcrossing with a recurrent parent and selection by MAS. The QTL in NILs can be handled as a single Mendelian factor, and this material can be used for mapping and positional cloning. In rice, a relatively large segment of a particular chromosome from the donor parent is substituted in the recurrent parental background to form Chromosome Segment Substitution Lines (CSSLs) whereas a very small chromosomal segment, containing the target QTLs or genes of a donor line, is substituted. .

Secondary mapping populations, such as CSSLs or NILs, are required to facilitate a more comprehensive analysis of target QTLs. Several rice primary and secondary mapping populations have been developed, and use of these will allow to perform the molecular cloning of target QTLs. In rice, QTLs for heading days have been well studied. Using a primary population derived from a cross between a Japonica variety, Nipponbare, and an Indica variety, Kasalath, many QTLs for heading days have been detected, and some have been mapped using a secondary population. So far, two QTLs, *Hd1* and *Hd6*, have been isolated by positional cloning.

Although artificially induced variations, such as mutants, have mainly been used for genetic and physiological studies in plant species, DNA markers have made it possible to the naturally-occurring allelic variations underlying complex traits and will become increasingly important for breeding. The use of wild relatives as donor parents would provide a more powerful way to exploit a wide range of allelic variation because these plants are adapted to specific environmental conditions. Because genetic variation gives rise to these different adaptations we might find a wider range of allelic variation in wild relatives than in cultivated species. Inter-specific crosses have already been used for QTL analyses in tomato, and rice.

There are many wild relatives in the rice *O. sativa* complex, AA genome, and seven species have been differentiated which have unique geographic distributions with differing environmental conditions. Using these wild species, it should be possible to identify and pull out the genetic information held in the rice genomic DNA for QTL analysis. QTL analysis, CSSLs covering whole rice chromosomes have been developed with wild relatives from *O. glaberrima* and *O. glumaepatula*. If QTL analysis can be used to locate traits related to adaptability and yield, and reproductive barriers such as hybrid sterility and growth abnormalities, it will give greater insight into how such crop species have evolved. The large degree of variation within the rice species will not only provide is with very important agronomic traits but also with very interesting information on plant biology.

12

PARTICIPATORY PLANT BREEDING

Participatory plant breeding (PPB) is a strategy used to strengthen on-farm conservation by encouraging farmers to continue to search, select, and manage local seed supply systems. The PPB process aims to consolidate the farmer's role in setting breeding goals and selecting diverse genetic materials. In so doing it offers skills and opportunities for farmers to search for new diversity and to select and exchange variable populations that match their local preferences and needs.

Participatory plant breeding has been advocated as a strategy for maintaining or enhancing the level of genetic diversity on-farm because it can be used to breed divergent cultivars for subtly different environments and for diverse end uses. Also PPB can add value to traditional landraces that would otherwise be lost from the system.

On the other hand, the success of PPB products may stem from the addition of just a few major genes, corresponding, for example, to pest resistance or plant height. It is also commonly assumed that the replacement of farmers' unique varieties with modern cultivars may affect the intraspecies genetic diversity by reducing the genotypic diversity and/or the allelic diversity within the cultivated crop.

Both participatory varietal selection (PVS) and PPB may have negative impacts on landrace diversity, because both methods are intended to change the structure of the local crop population to make it more productive and useful to farmers. It has been argued that successful PPB products may swamp a significant fraction of local diversity, leading to a short term gain in productivity, loss of local unimproved populations, and hence increased future vulnerability.

The creation of new varietal diversity by PPB, which has already been successfully used in difficult environments, is a powerful method that can be employed in high potential production systems. However, the impacts of PPB on the extent and distribution of local crop diversity are still poorly monitored and documented. The PPB programme in Nepal was designed to investigate:

(1) whether farmers' cultivars are being conserved,

(2) whether the PPB process encourages farmers to maintain processes that allow crops to evolve and change over time,

(3) whether PPB contributes to the enhancement and conservation of traditional varieties in situ and provides community benefits, and

(4) whether landraces can be conserved by improving preferred traits, hence making them more desirable in relation to modern cultivars.

The purpose of this chapter is to describe participatory methods that build local capacity to understand the extent and distribution of local crop diversity and the processes by which farmers maintain diversity on their farms. The farmers' needs, conservation issues, and diverse users' issues in the process of choosing parents and selecting variable populations.

A series of brainstorming sessions were organised in Nepal to integrate the PPB approach into the national crop breeding strategy. A primary stakeholder meeting was held in April 1998 to develop a PPB process for on-farm conservation and to agree upon the roles of formal and informal institutions. There are three essential components to PPB: (1) one parent is a locally adapted cultivar, (2) selection is decentralised in the target environment, and (3) farmers participate in the plant breeding process. Ceccarelli et al. argued that the most complete decentralisation could be achieved when farmers participate in the breeding process in their own fields.

Often plant breeders seek functional collaboration with farmers to obtain quick results. Community participation is central to on-farm conservation of agricultural biodiversity. The local capacity to enhance genetic diversity (i.e., to search, select, and exchange) is a key element of sustainable agriculture.

A broad PPB process was used in Nepal to develop local farmers' capacity to assess, select, and share local crop diversity, and to build sustainable partnerships between farming communities and researchers. The process can be divided into the following steps:

— Locate rice agroecosystems and identify interested communities.

— Organise "diversity fairs" for locating rice diversity, and collect germplasm and local knowledge.

— Understand and monitor local crop diversity through a community biodiversity register.

— Analyse options for adding benefits.

— Set breeding goals for PPB and parent selection with representative community participation.

— Agree on roles of farmers and researchers involved in the breeding process.

— Select for diversity.

— Strengthen farmers' seed systems for rapid diffusion.

In 1997 a "diversity fair" was organised at each site to raise community awareness of crop genetic resources, to locate genetic diversity and its custodians, and to promote the value of landrace diversity in the context of local food culture, market forces, and socioeconomic and agroecological considerations. The materials collected from the fairs were grown in farmers' fields as "diversity blocks".

The blocks were used to assess field performance and to analyse preferred and undesirable traits through participation from male and female farmers, representing all socioeconomic strata. The diversity block could also be used to select for appropriate parent plants and as a seed source for crossing programmes.

Understanding the value of local crop diversity is a key step prior to the initiation of participatory goal setting. After the diversity fair, local communities were motivated to keep a community biodiversity register (CBR). The CBR is a record, kept in a register book by community members, of all landraces in a community, including information on their custodians, passport data, and use value. It aims to monitor the level of diversity held by farmers in a community over time, as well as the number of households and area covered by each cultivar within the community.

Local communities are encouraged to use CBR information to understand factors influencing farmers' decision making on dynamic changes in local crop diversity and to develop their own on-farm conservation strategies.

Farming communities maintained a substantial level of rice diversity at the community level in both the Begnas middle hill eco-site and the highly accessible Bara site. More local than modern cultivars were maintained at the household level in both sites. The area of rice landraces in Bara was 17% of the total and in Kaski was 73%. Some landraces competed strongly with modern cultivars in certain niches and that these landraces could be promoted in similar areas without further improvement.

The dilemma prior to initiating PPB was deciding which rice varieties to include: all varieties, only those with high market demand, those maintained by only a few farmers in small areas, or those grown by many farmers in larger areas? Since no one knew which group was more important from a conservation perspective, and it was not possible to include all of the traditional cultivars, the team decided to divide the varieties into categories according to certain criteria.

There was no literature available to guide this research, so the team elected to divide the landraces into four groups, based on the average planted area and the average number of households growing each landrace. This design ensured that at least one variety from each group, representing a different use value, would be included in the crossing programme. The process used is given below:

— List existing local diversity.

— Categorise local crop diversity into four cells according to the planted area of each cultivar and the number of households growing each cultivar.

— Classify local crop diversity by use.

— Perform assessment analysis for preferred and unwanted traits using preference ranking and paired matrix ranking.

— Analyse potential benefits of local diversity and threats of genetic erosion.

A key element of the PPB programme is the choice of the local landrace parent. A consultative participatory process used to assess farmers' needs and the project goal highlighted the conflict of interest between choosing parents and setting breeding goals.

DIVERSITY AND SELECTION OF PARENTS

Participatory rural appraisals (PRAs) were carried out to (1) understand key factors affecting farmers' decisions to maintain local cultivars, including information on market development, and (2) ascertain key limiting factors to production systems reliant on local crop cultivars. To understand the relative importance of specific landraces, farmers' varieties were grouped into four broad categories using baseline survey data and CBR. This approach could be carried out at the species or variety level.

This type of broad distribution analysis helps us to understand, why some landraces are grown in a small area by many farmers, while others are grown in a small area by few farmers. It is very important for plant breeders to understand such distribution patterns and the underlying decisions made by farmers before starting to design a PPB programme to promote on-farm conservation.

Three landraces (Jetho budho, Ekle, and Mansara) from the commonly grown group were selected to address the needs of many households and to ensure that benefits would flow to a diverse range of people. Two landraces from the rare group (Birmaphool and Sano gurdi) were chosen. Aanga was chosen for its unique medicinal value and Gurdi was selected to represent good coarse varieties in each cell. Results indicated that few culturally important cultivars (e.g., Anadi in Kaski and Sathi in Bara) are grown by many households in small patches for household use. These landraces could be conserved per se; however, the population size of Anadi has increased in recent years with the linking of the project to markets and with sales being made under its brand name in departmental stores.

Local Seed System

Baniya et al. found that 96% of farming households in Begnas and Kochorwa villages are dependent upon informal seed sources. Seed flow occurs through farmers' social networks. These communities manage their rich rice diversity through bartering, gifts, borrowing seed or seedlings, and purchase community. These farmers are described as the "nodal" farmers of the community. They are perceived to be more knowledgeable than other farmers in relation to seed and production environment matters and as being "diversity minded".

Nodal farmers mainly belong to the resourceendowed farmer group, which is characterised by ownership of larger landholdings, more land parcels, and more livestock than less well-off farmer groups. Moreover, resource endowed farmers generally have a higher education level and participate more frequently in the local market. Some nodal farmers are women. Nodal farmers are spatially distributed within the community; together they can act as a conservation farming network and their farms are used as "field gene banks". They can be involved very effectively in community biodiversity registration and linked to development opportunities.

Nodal farmers can enhance farmer to farmer dissemination of genetic materials, be effective as resource persons for farmer to farmer training, and provide information on local crop diversity. Their expertise and knowledge can be effectively utilised in the development of training and extension materials on local cultivars, and they can also be involved in public awareness on agrobiodiversity.

The 13 segregating materials selected for PPB were distributed to nodal farmers and a few other interested farmers for growing F_2 or F_3 bulk of their choice. Field performance of the different populations was assessed during farm walks by researchers and farmers. In Begnas village, F_3 segregants of Mansara and Khumal-4 showed promising results in the upper areas (1000-1300 masl). The breeding goal was to incorporate the good eating quality and yield potential of Khumal-4 into Mansara without losing the latter's adaptive traits.

Three distinct population types were selected jointly by farmers and researchers and these were further advanced in farmers' fields. The spread of M x K-4 bulks is being monitored, and the selection history is being documented. The cross between Pusa basmati and Jetho budho also did well in Khola-ko-chewn environments. In Bara district, six large segregating populations were evaluated in farmers' fields, of which farmers selected the populations of Lajhi and IR62161-22-1-2-1-1 for lodging resistance and post harvest traits.

The participatory method used to divide local crop diversity into four categories is useful for understanding the extent and distribution of farmers' classification of local crop diversity. This is a simple method for teaching farmers about the concept of local common and rare traditional cultivars. It is also useful for analysing options for introducing other benefits such as increased community awareness and motivation, seed networks, new germplasm, market links, and PPB.

Social networks highlight the effectiveness of decentralised selection and informal seed systems. Nodal farmers' expertise in selecting and maintaining genetic materials can be effectively used in PPB. Furthermore, capacity building of nodal farmers in PPB may enhance crop diversity on a large scale.

Participatory methods such as biodiversity fairs, diversity blocks, and community biodiversity registers raise local awareness, strengthen local capacity for understanding the value of local crop diversity, and strengthen the roles of farmers and the informal sector in the process of local crop development. Such community participation helps the benefits of onfarm conservation to reach poor and biodiversity-based livelihoods. It unlikely that a single product of PPB will outcompete a large proportion of local crop diversity. The diverse agroecology existing in an area, combined with farmers' preferences for different cultivars with varying characteristics, should help to maintain biodiversity.

Participatory Plant Breeding (PPB) involves scientists, farmers, and others, such as consumers, extensionists, vendors, industry, and rural cooperatives in plant breeding research. It is termed "participatory" because users can have a research role in all major stages of the breeding and selection process. Such 'users' become co-researchers as they can: help set overall goals, determine specific breeding priorities, make crosses, screen germplasm entries in the pre-adaptive phases of research, take charge of adaptive testing and lead the subsequent seed multiplication and diffusion process. The fundamental rationale for PPB programme is that joint efforts can deliver more than when each actor works alone.

While some have cogently argued that commercial, private sector, plant breeding has long been client-driven, or 'participatory' under another name, the application of "PPB" to reach poor client groups, to breed for highstress, heterogeneous environments and to incorporate diverse traits to meet specific client preferences results in fundamental changes in the way plant genetic resources are managed by formal breeding programmes and farmers. It makes sense therefore, to analyse Participatory Plant Breeding as a new approach to germplasm development, especially in the public sector.

The CGIAR Programme on Participatory Research and Gender Analysis for Technology Development and Institutional Innovation (PRGA) currently has detailed documentation on 65 PPB programmes. Most of the cases, whether located in public sector or non-government (NGO) crop improvement programmes, were begun within the last 10 years. A lack of consensus about terminology is common when a new science is in its early stages, and Participatory Plant Breeding (PPB) is no exception. Terms commonly used interchangeably include: Collaborative Plant Breeding (CPB) Farmer Participatory Breeding (FPB); and Participatory Crop Improvement (PCI).

The latter is sometimes subdivided into two areas, one for work with stabilised materials, termed Participatory Varietal Selection (PVS) and another referring exclusively to work with variable or segregating materials, confusingly also sometimes termed PPB. All labels presently used describe broadly the same activities of what is a multi-faceted technical and organisational collaboration in plant breeding by scientists and users of their results.

PPB and Institutional Context

One of the most important differences among approaches to 'Participatory Plant Breeding' (PPB) is institutional, in the sense of the rules for behavior, the norms and values, and the incentives that govern how farmers and scientists will share the responsibility, the work and the benefits of a joint plant breeding effort. The key institutional difference lies in the built-in obligations which determine the locus of control or decisionmaking about the objectives of the plant breeding and the kind of results and data required to support these.

The distinguish two main institutional approaches: one when farmers join in breeding experiments which have been initiated by formal breeding programmes which term 'Formal-led PPB'; and another when scientists seek to support farmers own systems of breeding, varietal selection and seed maintenance, which call 'Farmer-led PPB.' The incentive structure, and rights and obligations which characterise these two approaches can be expressed in different types of organisational arrangements.

Formal-led PPB has certain unique institutional characteristics. Researchers run formal-led PPB programmes and invite farmer participation in the formal research. Researchers have an obligation and often a priority objective to feed information back to the formal research sector: this means that the scientific standards of replicability and validity of results must be met. There is the expectation that PPB will complement the formal sector research system, e.g. refining breeding strategies so that specific environments and varietal preferences are addressed, or re-orienting priorities.

Generally, formal-led PPB programmes also involve strong linkages to formal variety release and seed production systems. Finally, scientists involved in formal-led programmes are usually expected by the scientific community to extrapolate their methods, if not the varieties per se beyond the individual community with which they work. They often need to show what the advantages of PPB are compared to formal breeding approaches.

There are some distinguishing institutional characteristics of farmer-led PPB. Other professionals in farmer-led programmes are expected to facilitate a process in which farmers establish breeding objectives. Farmers bear the main responsibility for and, often, the costs of conducting experiments, selecting materials for seed multiplication, and dissemination of these. Researchers are expected to take a support role in this process.

Farmer-led PPB has the objective to provide varieties or populations which suit the specific local environment and local preferences and any broader applicability beyond local circumstances is fortuitous. Farmer-led PPB, with a few exceptions, tends to work for a specific client group or groups which have no obligation either to feedback information for wider geographical extrapolation, nor to feed products such as varieties into external formal release and seed systems.

It is important not to confuse the scale of a PPB effort with the institutional approach. The fact that PPB is carried out at the village or local scale does not mean that it is ipso facto farmer-led PPB. Since most PPB is still experimental and most initiatives in their early stages are conducted in a few sites, it is not yet clear whether there is an inherent difference in potential scale between the formal-led and farmer-led approaches.

Most farmer-led PPB is conducted at the community or local scale, and the locus of control over decision-making is local, but there are examples with broad geographical coverage involving several hundreds of communities or widely dispersed farmer groups. Farmer-led PPB might also produced broadly adapted varieties, although accomplishing this is not a priority. In contrast, formal-led PPB strategies are being implemented by large international or national public sector research bureaucracies, sometimes over large geographical areas, but these may deliberately build a mosaic of community-based and locally controlled varietal selection efforts which collectively service a large-scale production area.

Bio-social Environments of PPB

Two types of parameters have proved heuristic for characterising the environments in which PPB programmes are taking place. The first describes the type of agroecological environment in which PPB programmes develop. This has been plotted along a crop-specific scale ranging from high stress to low stress based on actual versus expected yields, coupled with an index for incidence of crop failure. The terms 'favored and unfavored' because these often mean "favored for staple cereal crops' but lay usage tends to ignore the notion of crop-specific comparative advantage, ie. a cool tropical highland environment is 'unfavored' for irrigated rice but highly favored for coffee, for example.

Agro-ecological environments potentially range from those which are primarily subsistence-oriented and highly unstable, implying that farmers' crop choices are governed by their own adaptive and preference needs, to systems in which crop production is very controlled and largely driven by urban consumer and/or commercial processor needs.

The second parameter suggests the broad economic environment of PPB, that is the degree "homogeneous demand versus heterogeneous demand" for varieties. Plotting was based on a nominal scale of 1' to '10' according to the 'leniency/ narrowness of varietal characteristics demanded by end-users' and the similarity/discordance between varieties used for home consumption and for sale. Contexts at the higher end tend to correspond to a high degree of homogeneity in product and often favor a narrow range of grain, taste, and cooking types. Such a high degree of uniformity/homogeneity is often associated with contexts where farmers are producing for highly specialised markets.

Some plant breeders consider PPB is most appropriate for environments which are high stress ('marginal') and where agriculture is low-input. Certainly, conventional breeding has been less effective in such difficult environments and in reaching farmers with few resources; so the rationale for testing 'participatory approaches' which are often site-specific, is a solid one. Analysis of actual PPB cases, however, shows a more complex picture. Not all PPB is concentrated in high stress environments with low input agriculture.

An unexpectedly large number of PPB programmes are being initiated in the intermediate areas where agroclimatic stress is less severe. On the whole these are cases where quality concerns, that is, meeting exigent end-user preferences' is defined as the paramount challenge. Precise plotting also shows that a significant amount of PPB work is now occurring in low stress areas where homogeneous end-user preferences are well-defined in the market. Two reasons explain most of the cases in these areas.

First, some of these PPB programmes aim to expand intra-crop varietal diversity in what have become relatively uniform farming areas.

Second, several others are run by NGOs or organised farmer groups with the primary goal of helping communities gain greater control over their breeding process or seed supply.

Goals of PPB

Over the last decade, PPB has been applied as a crop improvement strategy primarily in response to the need for impact in non-commercial crops and in very unpredictable, stressed production environments. Its successes in reaching highly diversified, very specialised and segmented markets have been less well publicised, albeit it is in this thematic area where the PPB challenges seem to be proliferating.

A range of other goals have also been defined within PPB programmes: for instance, enhancing biodiversity and germplasm conservation; developing adapted germplasm for especially disadvantaged user groups (eg. women, poor farmers); making breeding programmes more costefficient, particularly through decentralisation of programmes which target more niches. Close analysis of the set of PPB cases shows that some goals are explicit and often attained (for instance, production increase) while others are poorly articulated and usually not addressed, unless they are explicitly built into the research design. The trade-offs among goals is one of the areas where a good deal more structured or focused work needs to be pursued within the PPB field.

As partners usually have to accept trade-offs in reaching certain goals, it is important at the very beginning of a PPB collaboration for those concerned—scientists, farmers, development/NGO personnel-to discuss explicitly primary and secondary goals, and the minimal agreed-upon outcomes for which collaborators are aiming.

Participation and PPB

Participation (PPB) is a term used with a number of different connotations. However, it is essential to be clear about how to evaluate the separate dimensions of participation which together define what term its 'quality'. With respect to the 'quality of participation' in PPB, it is useful to identify three different dimensions: stage of participation; degree of participation; and actors' roles in participation.

The stage and degree of participation, together with the roles of the different actors need to be described in order to link different types of participation with different kinds of results. When researchers describe 'participation' in PPB programmes, they are generally referring to the stage of the breeding cycle at which farmers have been involved. It is usually fair to say that the earlier user participation occurs in a breeding process, the more opportunity users are given to influence the objectives, breeding strategy and final outcomes, but the extent to which users can realise this opportunity depends on the degree of participation.

A second dimension of participation is therefore, the degree to which farmers or other users who participate actually influence or make decisions about the process at any given stage. Descriptions of this dimension of participation in the cases studied are usually vague, reflecting a lack of clarity among PPB practitioners about the extent to which the degree of participation at any given stage of the breeding process can affect the end results. More time spent on participation cannot be assumed to be necessarily better quality participation, from either functional or empowerment perspectives. Poor women, especially, have enough to do without 'participating' in extra activities.

A third dimension of participation is the specific role taken either by the researchers, farmers or others. Role refers to the function performed: The role of actors in a participatory programme specifically refers to the functions they undertake, management role, information-giving role, or field labor provider.

Stage of Participation

After having agreed that a joint farmer-researcher collaboration in plant breeding is desirable (i.e. 'yes', do PPB) and having set the overall goals of the PPB (eg: biodiversity enhancement, farmer skill building, production increase), there are five stages which unroll, often cyclically:

1. Setting breeding targets
2. Generating (or accessing) variation through crossing (or using collections)
3. Selecting in segregating populations
4. Variety testing and characterisation
5. Interacting with seed systems (release, popularisation/marketing/diffusion, seed production, distribution).

The stage of farmer-researcher collaboration is one of the factors useful for comparing PPB cases. PPB may incorporate farmer input at various steps where it was not found in traditional breeding schemes. It may also significantly shuffle the order of these processes, e.g. breeders starting with 4 alongside farmers before solidifying stage 1, so that an iterative rather than linear research process is undertaken, with researchers, extensionists, farmers, traders or other kinds of participants taking different roles in each stage.

As many PPB efforts are linked to informal seed distribution, the need to understand existing seed systems can fit in with stage 1. Accompanying farmers in stages 4 and 5 can help breeders improve their role in stage 1 so that farmers may not subsequently need to be involved in 2 and 3 at all.

From examining stages of farmer involvement in the 65 cases, observe that farmer participation can usefully occur at various times, depending on the crop, parent materials, target region, researcher capacity to assimilate farmer criteria, farmer capacity to handle different types of materials, traits of interest, and scale of the breeding programme/number of materials to be screened. The stage at which farmer participation is first introduced to a conventional breeding programme can lead to change in the programme's objectives, or its breeding strategy, or even its organisation.

In many cases, the stages at which farmers participate or at which formal breeders participate changes as the programme develops, as the understanding of each others' skills and priorities increases. This often applies to the other realms of classification as well, i.e. degree of involvement and roles may evolve as the programme matures.

Degree of Participation

The degree of farmer participation is another dimension for classifying PPB. There, the degrees of participation were conceived of in the form of a wheel, which could evolve through time and according to the stage of involvement. The potential degrees of participation embraced the full range: from manipulative, passive, contract, consultative, collaborative, collegial through to farmer- or community-initiated.

In practice, three degrees of participation are generally found in PPB programmes: consultative, collaborative, and collegial. Consultative means that information is sought from farmers and, sometimes, from other clients of the breeding programme; collaborative means that there is task sharing between researchers and breeders, along lines determined by the formal research programme; collegial means that researchers support a farmer-initiated, farmer-managed programme which is accountable in a direct way to the farmers and other client groups with a stake in the results of the germplasm development.

Farmer-initiated work sometimes occurs at the later stages of formal-led PPB, usually at the very last stage of seed multiplication, distribution and popularisation. Farmer-initiated activities are also occasionally carried out within PPB programmes to support and strengthen farmers' local varietal selection, in situ conservation of germplasm, seed multiplication and distribution. Within the global review of PPB programmes, the most frequently observed degree of participation has been consultative and this takes place at the very first stage of defining breeding targets. If separate out the later stages, find that farmers are rarely involved in the PPB process in true sharing or decision-making roles at all. Nine cases were identified in which farmers worked with segregating materials.

Few of the cases analysed have experimented with collegial participation involving a significant devolution of responsibility to farmers. This may be because a good number of the cases are still testing approaches. There are as yet very few guidelines drawn from experience on the degree of devolution to farmers that can be achieved in a research programme which seeks to maintain certain standards of data quality which affect the replicability and validity of results. Programmes aimed more towards immediate developmental goals in specific locales might be expected to devolve more rapidly.

Roles of PPB

Farmers' and researchers' participation in PPB may have them taking on different roles or functions. In the cases analysed, the way in which researchers worked with farmers is not clearly described making it difficult to link stage of the breeding process in which participation is implemented, the degree of participation and the roles performed by researchers and farmers—with specific outcomes.

Based on the PPB cases analysed, the following roles taken on by farmers: management role: providing technical leadership, which involves a substantial technical contribution to the practical breeding process; management role: providing social organisational leadership, in which farmer-based institutions, eg. cooperatives, kinship-based networks serve as an organisational base by which PPB can be effected and/or scaled up; information-giving participation in which key insights, such as preferences, are used by others; and laborer or input supply role, neither of which imply any active participation in determining the outcome of the breeding process.

Finally, farmers play a key role, in providing germplasm to the breeding process. While formal breeding have used this farmer resource extensively, it has often been done without involving farmers specifically in the process of choosing germplasm, or in the subsequent process of evaluation and selection.

In some PPB cases, farmers have explicitly generated new base material for a shared breeding programme by making or facilitating crosses between chosen parents. Whether they are directly involved or whether farmer germplasm is used with direct farmer advice, the outcome of the programmes should recognise farmers' contributions when attributing any property rights to the finished materials.

13

Conservation of Plant Genetic Resources

Genetic diversity provides farmers and plant breeders with options to develop, through selection and breeding, new and more productive crops, that are resistant to virulent pests and diseases and adapted to changing environments. To meet the need for more food, it will be necessary to make better use of a broader range of the world's plant genetic diversity. Yet, genetic resources are disappearing at unprecedented rates. The reasons for this loss are many and include deforestation, developmental activities such as hydroelectric projects, road laying, urbanisation and changes in agricultural practices, and finally modern agriculture and introduction of new and uniform varieties. Over-grazing and changes in land-use pattern are taking heavy toll on diversity available in the wild species.

Urbanisation and changing life styles, globalisation and market economies are also contributing indirectly to the loss of diversity, particularly of minor and neglected crops. Such reductions have serious implications for food security in the long term. Global concern about loss of valuable genetic resources prompted international action. Programmes for conservation of plant genetic resources for food and agriculture were thus initiated and genebanks established in many countries. The main objective was to collect and maintain the genetic diversity in order to ensure its continued availability to meet the needs of different users.

The concept of germplasm conservation demands that collection methods initially capture maximum variation and subsequently, conservation and regeneration techniques minimise losses through time. To this effect, plant genetic resources (PGR) conservation activities comprise of collecting, conservation and management, identification of potentially valuable material by characterisation, and evaluation for subsequent use. Advances in biotechnology, especially in the area of in vitro culture techniques and molecular biology provide some important tools for improved conservation and management of plant genetic resources.

Germplasm Collecting

Collecting involves gathering samples of a species from populations in the field or natural habitats for conservation and subsequent use. The unit of collection may be botanic seeds or vegetative propagules, depending on the breeding system of the species. Collecting may be easy in species producing small botanic seeds in abundance. However, it becomes problematic when seeds are unavailable or non-viable due to: damage of plants by grazing or diseases; large and fleshy seeds that are difficult to transport; or where samples are not likely to remain viable during transportation due to remoteness of the collecting site from the genebank.

Advances in biotechnology provide useful solutions for collecting such problem species. For example, in coconut (Cocos nucifera), where the major difficulty for standard seed collection is the large size of the seeds, in vitro techniques have been developed that allow collecting of the relatively small zygotic embryos in the field and transporting them back in sterile conditions to the laboratory to inoculate and germinate them on a culture medium.

In cacao (Theobroma cocoa), where collecting germplasm in the field is limited by the rapid deterioration of samples during transit as the seeds do not withstand desiccation, some researchers reported a simple in vitro method that involved collecting shoot nodal cuttings, followed by sterilisation and inoculation of tissue into prepared culture vials containing semi-solid medium. In vitro collecting methods were also developed for a range of other species including oil palm, forage grasses, banana, coffee, grape, Prunus and Citrus spp.

There are two approaches for conservation of plant genetic resources, namely in situ and ex situ. In situ conservation involves maintaining genetic resources in the natural habitats where they occur, whether as wild and uncultivated plant communities or crop cultivars in farmers' fields as components of the traditional agricultural systems. Ex situ conservation on the other hand, involves conservation outside the native habitat and is generally used to safeguard populations in danger of destruction, replacement or deterioration.

Approaches to ex situ conservation include methods like seed storage, field genebanks and botanical gardens. DNA and pollen storage also contribute indirectly to ex situ conservation of PGR. Among the various ex situ conservation methods, seed storage is the most convenient for long-term conservation of plant genetic resources. This involves desiccation of seeds to low moisture contents and storage at low temperatures. However, there are a large number of important tropical and sub-tropical tree species which produce recalcitrant seeds that quickly lose viability and do not survive desiccation, hence conventional seed storage strategies are not possible.

There are also a number of other important crop species that are sterile or do not easily produce seeds, or seed is highly heterozygous and clonal propagation is preferred to conserve elite genotypes. Examples of crop species in this category are banana, sweet potato, sugarcane, cassava, yam, potato, and taro. These species are usually conserved in field genebanks. Although field genebanks provide easy access to conserved material for use, they run the risk of destruction by natural calamities, pests and diseases. For this reason, safety duplicates of the living collections are established using alternate strategies of conservation and it is in this area that biotechnology contributed significantly by providing complementary in vitro conservation options through tissue culture techniques.

In vitro conservation also offers other distinct advantages. For example, the material can be maintained in a pathogen-tested state, thereby facilitating safer distribution. Further, the cultures are not subjected to environmental disturbances. Several in vitro techniques have been developed for storage of vegetatively propagated and recalcitrant seed producing species. In general, they fall under two categories: (i) slow growth procedures, where germplasm accessions are kept as sterile plant tissues or plantlets on nutrient gels; and (ii) cryopreservation where plant material is stored in liquid nitrogen. Slow growth procedures provide short- and medium-term storage options, while cryopreservation enables long-term storage of the plant material.

Slow Growth

Slow growth procedures allow clonal plant material to be held for 1-15 years under tissue culture conditions with periodic sub-culturing, depending on species. There are several methods by which slow growth can be maintained. In most cases, a low temperature often in combination with low light intensity or even darkness is used to limit growth. Temperatures in the range of 0-5°C are employed with cold tolerant species, but for tropical species which are generally sensitive to cold, temperatures between 15° and 20°C are used. It is also possible to limit growth by modifying the culture medium, mainly by reducing the sugar and/or mineral elements concentration and reduction of oxygen level available to cultures by covering explants with a layer of liquid medium or mineral oil.

Regeneration and successful propagation of genetically stable seedlings from cultures are prerequisites for any in vitro conservation effort. Protocols for clonal multiplication are well established for several species. Generally, organised cultures such as shoots are used for slow growth storage since undifferentiated tissues such as callus are more vulnerable to somaclonal variation. Although slow growth procedures have been developed for a wide range of species, they are routinely used for conservation of genetic resources of only a few species including Musa spp., potato, sweet potato, cassava, yam, Allium spp. and temperate tree species. About 37,600 accessions are reportedly conserved by in vitro techniques in genebanks, worldwide.

Cryopreservation

Cryopreservation involves storage of plant material at ultra-low temperatures in liquid nitrogen (–196°C). At this temperature, cell division and metabolic activities remain suspended and the material can be stored without changes for long periods of time. Cryopreservation is the only available method for long-term conservation of vegetatively propagated plant germplasm. The choice of material includes cells, protoplasts, shoot apices, somatic embryos, seed or excised zygotic embryos. Cryopreservation requires limited space, protects material from contamination, involves very little maintenance and is considered to be a cost-effective option.

The techniques for cryopreservation currently in use are quite varied and include the older classical techniques based on freeze-induced dehydration of cells as well as newer techniques based on vitrification. In classical techniques, tissues are cooled slowly at a controlled rate (usually 0.1-4°C/min) down to about -40°C, followed by rapid immersion of samples in liquid nitrogen. Slow freezing is carried out using a programmable freezing apparatus. Cryoprotectants are added to the freezing mixtures to maintain membrane integrity and increase osmotic potential of the external medium.

Classical cryopreservation procedures have been successfully applied to undifferentiated culture systems such as cell suspensions and calluses. However, in case of differentiated structures, they have been employed for storage of apices or embryonic axes of only cold-tolerant species, and their utilisation for tropical species has been limited. Vitrification-based procedures involve removal of most or all freezable water by physical or osmotic dehydration of explants, followed by ultra-rapid freezing which results in vitrification of intracellular solutes, i.e. formation of an amorphous glassy structure without occurrence of ice crystals which are detrimental to cellular structural integrity.

These techniques are more appropriate for complex organs like embryos and shoot apices; they are also less complex and do not require a programmable freezer, hence are suited for use in any laboratory with basic facilities for tissue culture. There are seven vitrification-based procedures in use for cryopreservation:

1. encapsulation-dehydration,
2. vitrification,
3. encapsulation-vitrification,
4. desiccation,
5. pregrowth,
6. pregrowth-desiccation, and
7. droplet freezing.

With the advent of these new cryogenic procedures, especially vitrification, encapsulation-vitrification and encapsulationdehydration, the number of species cryopreserved has increased significantly in recent years. In general, cryopreservation is well established for vegetatively propagated species. However, it is much less advanced for recalcitrant seed species due to some of their characteristics, including their very high sensitivity to desiccation, structural complexity and heterogeneity in terms of developmental stage and water content at maturity. The new cryopreservation techniques have been successfully applied for more than 80 species and they are under development or vigorous testing for several other species. However, examples of their routine use for long-term conservation are still limited only to oil palm and potato.

CHARACTERISATION OF DIVERSITY

The ability to identify genetic variation is indispensable to effective management and use of genetic resources. Traditionally, diversity is assessed by measuring variation in phenotypic traits such as flower colour, growth habit or quantitative agronomic traits like yield potential, stress tolerance, etc., which are of direct interest to users. This approach has certain limitations: genetic information provided by morphological characters is often limited and expression of quantitative traits is subjected to strong environmental influence.

In the 1960s, biochemical methods based on seed protein and enzyme electrophoresis were introduced. which proved particularly useful in analysis of genetic diversity as they reveal differences between seed storage proteins or enzymes encoded by different alleles at one (allozymes) or more gene loci (isozymes). Use of biochemical methods eliminates the environmental influence; however, their usefulness is limited due to their inability to detect low levels of variation: DNA-based techniques introduced over the past two decades have potential to identify polymorphisms represented by differences in DNA sequences.

These methods are being used as complementary strategies to traditional approaches for assessment of genetic diversity, the major advantage being that they analyse the variation at the DNA level itself, excluding all environmental influences. The analysis can be performed at any growth stage using any plant part and it requires only small amounts of material. Following the advances in molecular biology in the last decade, a variety of different methods have been developed for analysis of genetic diversity. These methods differ with respect to technical requirements, level of polymorphism detected, reproducibility and cost.

Molecular methods used for detecting DNA sequence variation are generally based on the use of restriction enzymes that recognise and cut specific short sequences of DNA (e.g., Restriction Fragment Length Polymorphism, RFLP) or polymerase chain reaction (PCR), which involves amplification of target DNA sequences using short oligonucleotide primers. PCRbased techniques such as Random Amplified Polymorphic DNAs (RAPDs), Amplified Fragment Length Polymorphisms (AFLPs) and Simple Sequence Repeats (SSRs, microsatellites) have proved especially important in diversity studies.

Newer and more powerful molecular techniques that detect variation at specific gene loci, which can be automated for high throughput of samples, are becoming available, permitting precise and versatile analyses of genetic variation. Nevertheless, it is difficult to compare the different techniques and determine which one is best and for what purpose because each method has its advantages and disadvantages. The appropriateness of individual marker systems also varies depending on the objective of study and the properties of the species.

Molecular methods are increasingly playing an important role in conservation and use of plant genetic resources. Specific areas in which molecular marker techniques have been used are: developing sampling strategies and identification of gaps in the collections to plan for future collecting and acquisition, and managing conserved germplasm—including identification of redundancies, development of core collections, fingerprinting, identification of genetic contamination and quantification of genetic drifts/shifts.

Managing Genetic Resources

Molecular techniques proved useful in a number of ways to improve the conservation and management of PGR. In particular, genetic diversity data provides information on gaps in terms of coverage in gene pools as well as redundancies, i.e., material with similar characteristics that wastes resources through increased cost of management. For example, RAPD analysis in Brassica oleracea revealed that 14 phenotypically uniform accessions could be reduced to 4 groups with minimal loss of genetic variation.

Intraaccession variation was determined by AFLP markers in ex situ conserved barley, and results were used to evaluate the efficiency of splitting heterogeneous accessions into distinct lines in order to avoid the negative effects of selection and genetic drift during regeneration. Molecular markers have been used to identify groups from which core collection accessions can be selected or to monitor the effectiveness of one or the other strategy in capturing genetic diversity found in the whole collection. For example, molecular markers were used to assist the assembly of a core collection for cultivated potato species Solanum phureja.

Using RAPD markers, some researchers found no significant differences in genetic diversity between core collection and a chosen set of accessions, randomly chosen from the whole collection of common bean germplasm, indicating that the method used to develop the core collection had not been very efficient in capturing diversity. Molecular markers have been employed for fingerprinting, verification of accession identity and genetic contamination. For example, microsatellites were used to distinguish different cultivars of grapevine, and to compare landraces and develop unique DNA profiles of soybean genotypes.

Variation within species has also been studied to explore geographic or ecological patterns of distribution of diversity in many different crops and their wild relatives that include wild bean (Phaseolus vulgaris), banana, mango, bambara groundnut, vetch, Cicer spp., sorghum, sweet potato, tea and chicory. In Lycopersicon, microsatellites analysis revealed high level of polymorphism within and among species, which was highly correlated with the respective mating systems, cross-pollinating species having a significantly higher gene diversity compared to self-pollinating species.

In Situ Conservation

Diversity studies using molecular markers have also assisted in developing in situ conservation strategies. The advent of molecular genetic markers made it easy to discriminate between wanted and unwanted agronomic genes in segregating populations. If linkages are established between a heritable agronomic trait and a genetic marker, markers can be used to identify the location of genes. Such linkages allow direct selection for the trait using marker assisted selection in a backcrossing programme. Molecular markers which densely cover an entire crop genome can be applied to develop a molecular map for a crop, which could be used to determine linkage between a specific molecular marker and a strongly heritable trait This holds great promise for breeding programmes as many traits are difficult to select for directly from breeding populations.

Germ Plasm Exchange

The ability to store and exchange healthy germplasm is fundamental to effective conservation and use of plant genetic resources. Infected propagules are known to deteriorate quickly in sub-optimal storage conditions, therefore it is important to store only healthy samples in the genebanks. Further, because planting material is an international commodity used for exchange of germplasm, infected seeds inadvertently serve as means of introducing plant pathogens into new areas. For the above reasons, seed health tests and phytosanitary regulations are generally incorporated as part of the genebank activities.

Prior to the 1970s, seed health tests were primarily directed towards fungi and relied on incubation or grow-out tests for detection of pathogens on the seed. Such tests are labour intensive, time consuming and in some cases seed require high levels of infection before pathogens can be detected. The need to use representative sample sizes in the traditional growout methods to detect pathogen contamination also poses a problem when only limited amount of seed is available for genetic conservation. Also, pathogens that are symptomless such as viruses and viroids pose great risk in exchange of germplasm as they cannot be detected by traditional methods of bioassay.

The problems are more severe for crops propagated by vegetative means, as viruses and viroids are transmitted very efficiently through vegetative propagules. The demand to store and exchange seeds of high quality and for greater sensitivity in detecting seed-borne pathogens including viruses and bacteria and shortturnaround times for seed testing have led many genebanks and seed health testing laboratories develop or adopt new technologies. Barker and Torrance recently reviewed the increasing role and future potential of biotechnology in detection of pathogens from germplasm collections. In lettuce, for example, the method for detection of lettuce mosaic virus (LMV) changed since 1983 and a serological test—enzymelinked immunosorbent assay (ELISA) using specific LMV antiserum was developed replacing the standard seed assays which require a grow-out and evaluation of 30,000 lettuce seedlings.

ELISA is not only more efficient, but it has proved to be sensitive in finding low levels of seedborne LMV that could devastate lettuce production. Immunofluoresence is an improved technology that is routinely used for screening seed samples for bacterial pathogens. The PCR technique used to amplify small amounts of DNA, also improves the efficiency of seed health tests, besides sensitivity and/or specificity. For example, in watermelon, PCR has been developed for detecting bacterial fruit blotch pathogen as an alternative to growout tests requiring 10,000 seedlings.

DNA primers have been developed for several important seedborne pathogens, and PCR techniques are in routine use for the detection of blackleg of crucifers, halo blight of beans and bacterial canker (CMM) of tomato. Diagnostic tools that use oligonucleotide-Probe (DOP) hybridization are now available to detect coconut cadang-cadang viroid. Detection of pathogens has to be followed by elimination for safe exchange or conservation of the material. Apical meristem culture has been used to successfully eliminate many viruses from a variety of plant species. Meristem culture is usually combined with heat treatment for better results. Ng described tissue culture techniques for both disease elimination and micropropagation in cassava.

Postman reported successful elimination of viruses from 400 pear cultivars conserved at USDA-ARS National Clonal Repository. Thermotherapy and meristem-tip culture were used for elimination of viruses from cassava collections in the in vitro genebank at the International Center for Tropical Agriculture (CIAT) (IPGRI/CIAT 1994). Other reports of successful elimination of viruses from important crop species include cassava mosaic virus from cassava, exocortis viroid from Citrus species, potato virus S from potato and sweet potato yellow dwarf virus from sweet potato plants.

Exchange of germplasm as in vitro cultures offers considerable advantages like reduced volume and weight as well as the improved health status of the cultures. Crop species for which in vitro germplasm exchange is a routine include Musa spp., potato, yam and cassava. Cultures are generally transported as shoot cultures on standard media, or as microtubers in case of tuber crops like potato, sweet potato and yam.

In banana, Rao et al. (1993) reported the use of shoot tips encapsulated in alginate beads and in yam and potato, Hasan and Takagi (1995) described the use of encapsulated nodal segments as an alternative form for distribution of germplasm. These techniques, besides reducing the weight and volume of the material to be distributed, retain the sterile conditions and simplify the process of regeneration upon receipt of the material.

Usage of Germplasm

Biotechnological advances have offered new approaches to

overcome challenges for effective utilisation and enhancement of crop genetic resources. For instance, a number of techniques have been developed to overcome problems of sexual incompatibility that lead to hybrid sterility or lack of genetic recombination in wide crosses involving distant wild relatives and cultivated species.

Embryo Rescue

In embryo rescue, an otherwise non-viable hybrid embryo is transferred to a culture medium where viable plants may be regenerated and backcrossed to the cultivated species to introduce the desired genetic trait. There are several examples of application of embryo rescue for a wide range of agronomically important species. Early examples of successful wide crosses using embryo rescue technique are: interspecific hybrids between Lycopersicon esculentum and L. peruvianum, and Medicago sativa and M. rupestris.

Hybrids have been obtained between Arachis hypogaea and the incompatible species A. paraguariensis and A. appressipila, both resistant to early leaf spot, using embryo rescue and tissue culture techniques. A combination of ovary culture and embryo rescue was used to develop fertile hybrid plants from the intergeneric cross between Brassica napus and

Sinapsis alba, which has many desirable traits like resistance or tolerance to all major insect pests of brasicca crops, tolerance to high temperatures and drought besides being shatter resistant. Embryo culture proved to be a useful tool to overcome postzygotic incompatibility in different Helianthus spp. and facilitated transfer of resistance to broomrape (Orobanche cernua). Recently, resistance to late blight from Solanum pinnatisectum was introgressed into S. tuberosum using embryo rescues and double pollination.

Somatic Hybridization

Protoplast fusion and somatic hybridization provide an alternative way for transfer of traits between distantly related species. It has been particularly useful in breeding programmes to transfer beneficial characteristics from wild and weedy plants to the cultivated crop species, breaking the barrier for gene transfer. Protoplast surfaces bear strong negative charges, and intact protoplasts in suspension repel each other. Hence fusion is accomplished by addition of calcium ions or polyethylene glycol (PEG) or using electric fields. Successful gene transfer via protoplasm fusion depends on the ability to regenerate a mature plant from the fusion product.

There are several examples where pre-zygotic sexual incompatibilities are overcome using somatic protoplast fusion and plant regeneration from the heterokaryons formed by interspecies protoplast fusions. Thus, intertribal somatic hybrids were produced between Brassica napus and Thlaspi perfoliatum with high contents of nervonic acid. Recently, protoplast fusion between B. napus and related wild species Orychophragmus violaceus enabled the transfer of genes for desirable fatty acid composition into B. napus. Porteresia is a halophytic species which can withstand total submergence in seawater and taxonomically related to rice (Oryza sativa).

Pre-zygotic incompatibilities have resulted in the species being recalcitrant to sexual hybridization with O. sativa. Production of heterokaryons by the fusion of mesophyll protoplasts of P. coarctata and cell suspensions derived from protoplasts of O. sativa has been achieved and somatic hybrid plants have been produced after regeneration. Somatic hybrids were produced and late blight resistance was successfully transferred from Solanum nigrum into S. tuberosum.

Bibliography

Ayad, W G, Hadgkin T, Jaradat A and Rao V R, "Molecular genetic techniques for plant genetic resources", *In: Proceedings of Meeting held at Rome, Italy, Rome, Italy: International Plant Genetic Resources Institute*, 1997.

Barton, J H and Siebeck W E., "Material transfer agreements in genetic resources exchange: the case of the International Research Centres", *Rome, Italy: International Plant Genetic Resources Institute*, 1994.

Duvick, D N., "Plant breeding and biotechnology for meeting future food needs", *In: Population and food in the early twenty-first century: meeting future food demands of an increasing population*, Washington DC, USA: *International Food Policy Research Institute*, 1995.

Fraley, Robert T. Global Focus Spring, "Biotechnology and sustainable agriculture", *Biotechnology for Biological Control of Pests and Vectors*, FL: CRC Press, 1991.

Gomez, O, Depestre T and Casanova A., "Germoplasma/Germplasm/ Germoplasme horticultural problems in the tropics: breeding for less favoured environments", *Proceedings of the Twenty-eighth Annual Meeting of the Caribbean Food Crops Society, Santo Domingo*, Dominican Republic, 1993.

Hauptli, Holly, Katz, David, Thomas, Bruce R., Goodman, Robert M., "Biotechnology and crop breeding for sustainable agriculture", *In Sustainable Agricultural Systems: proceedings of the International Conference on Sustainable Agricultural Systems held at Ohio State University*, Columbus, Ohio, 1990.

McCuen, Gary E., "Organic Farming and Biotechnology", *Poison in Your Food*, McCuen Publications, 1991.

Murray, D R, *Advanced methods in plant breeding and biotechnology*, Wallingford, UK: CAB International, 1991.

Villalobos V M and Engelmann F., "*Ex situ* conservation of plant germplasm using biotechnology", *World Journal of Microbiology and Biotechnology*, 1995.

_______________, "Impacts of biotechnology in International Agriculture and forestry", *The impact of plant molecular genetics, Boston*, USA: Birkhauser Boston, 1996.

Whitten, M. J., Oakeshott J. G., "Opportunities for modern biotechnology in control of insect pests and weeds, with special reference to developing countries." *FAO Plant Protection Bulletin*, 1991.

Index